The Mink

for Jane

... and Sula

The
MINK

Nigel Dunstone

Illustrated by
John Davies

T & A D
POYSER
NATURAL
HISTORY

© T & A D Poyser Ltd

First published in 1993 by T & A D Poyser Ltd
24–28 Oval Road, London NW1 7DX

All rights reserved. No part of this book may be reproduced, stored in a retrieval system, or transmitted in any form or by any means, electronic, mechanical, photocopying or otherwise, without the permission of the publisher

Typeset by Phoenix Photosetting, Chatham, Kent
Printed and bound in Great Britain by
Mackays of Chatham PLC, Chatham, Kent

A catalogue record for this book
is available from the British Library

ISBN 0–85661–080–1

Contents

List of colour plates	vi
Acknowledgements	viii
CHAPTER 1 *Introduction*	1
CHAPTER 2 *What is a Mink?*	4
CHAPTER 3 *Finding Mink: Tracks and Signs*	14
CHAPTER 4 *Phylogeny, Distribution and Status*	18
CHAPTER 5 *Adaptations to Habit and Habitat*	34
CHAPTER 6 *Food and Foraging*	62
CHAPTER 7 *Lifestyle*	100
CHAPTER 8 *Sex and Society*	140
CHAPTER 9 *Population Biology*	151
CHAPTER 10 *The Fur Trade*	163
CHAPTER 11 *Interactions with Man and other Animals*	187
References	206
Species Index	217
Index	223

The colour plate section can be found between pages 120–121.

List of Colour Plates

The colour plate section can be found between pages 120–121.

1. Mink carrying perch to den. Photo: Nigel Dunstone.
2. Mink carrying woodmouse. Photo: Nigel Dunstone.
3. Mink investigating horse clams. Photo: Dave Hatler.
4. Mink in alert posture. Photo: Nigel Dunstone.
5. Young mink sitting on rock den. Photo: Nigel Dunstone.
6. Mink peering at fish prey. Photo: Nigel Dunstone.
7. Adult mink in alert posture. Photo: Nigel Dunstone.
8. Prey size selection experiment using shore crabs. Photo: Nigel Dunstone.
9. Ventral spot pattern used in identifying individual mink. Photo: Nigel Dunstone.
10. Typical riverside habitat favoured by mink. Photo: Nigel Dunstone.
11. Mink habitat in northern Spain. Photo: Nigel Dunstone.
12. Coastal habitat utilized by mink in southern Scotland. Photo: Nigel Dunstone.
13. Pastel coloured mink. Photo: Nigel Dunstone.
14. Tree root den. Photo: Nigel Dunstone.

List of Colour Plates

15. Breeding den with accumulation of scats. Photo: Nigel Dunstone.
16. Radio-collared mink. Photo: Nigel Dunstone.
17. Taking measurements from an anaesthetized wild mink. Photo: Nigel Dunstone.
18. Barn used by mink as a winter den. Photo: Nigel Dunstone.
19. 'Asking for trouble?' Free range chickens on a mink infested river in North Yorkshire. Photo: Nigel Dunstone.
20. Gamekeepers gibbet with mink, stoat and weasel. Photo: Rob Strachan.
21. Young mink kit at approximately six weeks. Photo: Dave Hatler.
22. Newly born mink kit. Photo: Dave Hatler.
23. Severe wound on the neck of a female, a result of a mating fight. Photo: Dave Hatler.
24. Pattern of moult in a coastal mink from Vancouver Island. Photo: Dave Hatler.

Acknowledgements

I express my sincere gratitude for the considerable help and assistance I have received from many colleagues and friends during the years that I have worked on mink and latterly in the production of this book. Ernest Neal encouraged me to put pen to paper, and I thank him for his guidance and support during its lengthy gestation.

For my introduction to the world of mink biology I am grateful to Drs. Trevor Poole and Willie Sinclair, who in guiding my early days of research set the scene for what has shaped the past 15 years. During that time I have had the pleasure to collaborate with a succession of research students, and I particularly thank, Andy Clements, Moira Owen, Sharon Davies and Jonathan Gregory for helping unravel the mink myth by means of their carefully conducted laboratory studies. For field-based research I am grateful for the diligent observations made by Johnny Birks and Mark Ireland. I am also greatly indebted to Sharon Davies and Mark Ireland for allowing me to use some of their unpublished observations. I would like to express my gratitude to Dave Hatler for allowing me to share some of his wonderful observations of mink made on Vancouver Island during the mid 1970s, and for providing me with a wealth of photographic material which formed the basis of many of the illustrations presented.

None of my studies on the behaviour and ecology of mink could have been conducted without the assistance provided by Brian Finlay of Ross Farm, and Tommy and Peter Harrison—my 'tame' mink farmers and friends for many years. In the University of Durham I express my thanks to Paul Loftus, for devotedly preparing countless kilogrammes of mink scats, Peter Hunter and his staff for maintaining my mink colony, Dave Hutchinson for photographic assistance and Jean Mather for typing parts of the manuscript, and to the remainder of my colleagues for putting up with the often odoriferous mink!

Above all, I wish to record my sincerest thanks to Jane O'Sullivan for correcting my thinking as well as my grammar, for her tireless assistance, for cogent argument, and for moral support during the course of the preparation of this text.

CHAPTER 1

Introduction

Venus and Satan have never been so curiously confused as in the popular image of the mink. For many, the name conjures up luxury and romantic extravagance, the stuff of dreams and Hollywood. For others the mink is the devil's own work, a ferocious and passionless destroyer of all creatures, wild and domestic, that fall within its grasp. For such a small and secretive animal to carry two such claims to notoriety is remarkable indeed. It suggests a story worth the telling.

The mink's economic role as a valuable fur bearer and a reputed pest, has drawn much scientific attention, while the animal's private life has rewarded and sustained the interest of biologists. Often it is an animal's specialized adaptations to habitat that make them interesting to study. By contrast, the mink seems to be the supreme generalist, capable of thriving in diverse habitats with remarkable versatility. While its habitat requirements may be loose, the mink experiences strong selective pressure from intraspecific competition. This has lead to an interesting array of adaptations in its behaviour and reproductive biology, to allow each individual to increase the probability that it will be represented genetically in the next generation. The concerns of the male and the female differ almost entirely in this regard, and it is fascinating to examine how each coerces or confounds the other's efforts.

Superficially, mink share the general mould of all mustelids: an elongated, flexible and streamlined body, with short limbs and a small, pointed head. A form so successful that it had changed very little during 50 million years of evolution. In the Oligocene epoch the ancestral mustelids (e.g. *Plesictis*) stalked their prey looking much as they do today. Their descendants now occupy a great diversity of habitats, and range from the delicate weasels to the formidable giant otters of South America. Among them, the mink occupies the middle ground, intermediate in size and habit between the terrestrial species, such as the weasel and polecat, and the aquatic otters.

Charles Darwin was quick to recognize the minks equivocal position in defence of his evolutionary theory, which required that animals could evolve from one form to another in gradual steps, while 'each (intermediate grade) must be well adapted to its place in nature'.

> It has been asked by opponents of such views as I hold, how for instance, could a land carnivorous animal have been converted into one with aquatic habits; for how could the animal in its transitional state have subsisted? . . . Look at the *Mustela vison* (mink) of North America.
>
> *Origin of Species* Darwin (1859)

Much of my personal research as documented in the following chapters has been directed at assessing the semi-aquatic mink's adaptations and determining the ecological niche occupied by this predator. In fact, the American mink did not long remain an exotic scientific abstraction. Its coveted fur coat caused it to be transported by man extensively throughout the northern latitudes, and its versatility has enabled the inevitable escapees to establish thriving feral populations in most of those regions.

Without dependence on any particular prey type, the mink can minimize its direct competition with resident predators. Thus it has spread throughout Europe and the subarctic with few apparent barriers and, despite its reputation, with remarkably little impact on the ecology of the indigenous fauna.

It is none the less easy to see how the mink has become one of the most despised creatures in the countryside. Many country people, who have a conflict of interests, tend to view all predators as undesirable vermin. As an alien in Europe, and unknown quantity, the mink attracts particular suspicion. Its spread has suspiciously coincided with the decline of many native species who may qualify as its prey. The role of the concurrent and unprecedented rate of habitat destruction by man may be less obvious to the casual observer. Already framed as the villain, the mink's aggressiveness and sharp teeth do not endear it to impartial handlers. Even its versatility, often exaggerated, earns the perverse admiration conferred on a worthy adversary:

> Make no mistake the mink is a super stoat . . . it can climb trees in a way that compares with a squirrel, swim underwater at speeds fast enough to catch a trout, and it is very nippy for a short distance on land. In addition to these all round Daley Thompson abilities, the mink is bold to the point of foolhardiness.
>
> *Shooting Times and Country Magazine* (9 May 1985)

While the mink is no model of benevolence, many years of research in the field have revealed no evidence of the ecological disasters so often predicted in the past. To what extent is the reputation of the feral mink deserved? I hope this book may make the reader a fair judge. At least, I hope it will help put an end to the gross misrepresentation of the animal in the popular press.

I have referred specifically to the American mink, *Mustela vison*, but there are in fact two species who commonly carry the name of 'mink'. The second is *Mustela lutreola*, the European mink. It ranges across the northern reaches of Eurasia, from Siberia to Spain, although it is now rare or absent in many of its traditional realms. Visually the two species are very similar, but the relationship between them, and their distinction from related species such as polecats, has a tenuous scientific basis, as will be discussed in Chapter 3. For the purpose of this book the colloquial grouping is retained and I have attempted to present the story of both species. However, while a great body of knowledge has amassed on the American mink, very little systematic research has been done on its European cousin. For much of its ecology and physiology it may be assumed to resemble *M. vison*, and wherever information is available on the European mink, it is presented in its appropriate context throughout the book. Unless specified, the text refers to the American mink.

A readable book on mink has been needed for some time. There can be few animals that have been maligned so frequently and misrepresented so consistently. For myself, the mink has been the subject of intensive research over nearly two decades, and has led me on a scientific safari from the laboratory to the field and back again. Like a sleuth on the trail of a felon, my scientific wanderings and those of my research colleagues over the years have helped bring to light many aspects of its biology. Yet many questions remain unanswered, and some will be for ever unanswerable. I hope this book might stimulate others to fill in the gaps in our knowledge of the mink.

CHAPTER 2

What is a Mink?

GENERAL CHARACTERISTICS

The mink is a medium-sized carnivore, smaller than an otter, yet larger than a stoat, it is approximately 30 cm long not including the tail. Like all members of the weasel family (Mustelidae) to which it belongs, the mink's body is elongated with relatively short limbs (Fig. 2.1). Each of the limbs bears five digits with a little webbing between the toes. Each digit terminates in a compressed, curved claw which is non-retractile. The tail is long and slightly bushy, approximately a third of the body length. The head is flattened and wedge-shaped tapering to the muzzle, giving the face a somewhat pointed appearance. The jaws are short and powerful, befitting a predator.

THE COAT

A luxuriant and lustrous fur coat is synonymous with the name of mink. Aesthetic appeal aside, the mink's thick waterproof coat is highly functional for its semi-aquatic habits in the cold northern temperate zone. The range of colours commercially available is not entirely nature's doing; wild native American mink and European mink are uniformly dark brown, almost black, in colour. Feral mink may vary in colour, but after a few generations breed back to the dark brown 'wild type' colour. The coat quality of feral animals is rarely that of ranched mink.

Selective breeding of ranch mink produced a plethora of coat colours (see Chapter 10). The genes for these colour varieties are usually recessive, but they remain in the feral population and aberrant colours occasionally turn up in the wild. In Devon, Birks (1986) found 3% of the animals he trapped to be pale silver-grey (called 'Silverblu' by the fur trade). In south-west Scotland such animals comprised 9% of the population, with a further 2% of a pale brown variety (referred to as 'Pastel' or 'Sapphire'). Linn & Stevenson (1980) noted that the pelts of feral animals seem to be darker than those of the earlier

FIGURE 2.1 *Relative body sizes of representative mustelids. (a) Otter, (b) mink, (c) stoat, (d) weasel.*

generations found breeding in the wild. White mink are also occasionally found in feral populations (Fairley 1980).

Native American and feral mink usually have a number of white spots on the underside of the body, particularly on the chin, lower lip, abdomen and groin area. A white tip to the tail has also been occasionally recorded (Fairley 1980). Figure 2.2 shows the extent of variation in ventral white patches in a sample of feral mink trapped in Ireland. The distribution of white spots and blotches does not change substantially as a mink ages, although there may be minor changes in outline (Chanin 1983). Given that these white spots are relatively consistent during the lifetime of an animal, they can be useful in recognizing individuals in the population. Most studies that have involved long-term, live-trapping of mink have used this technique.

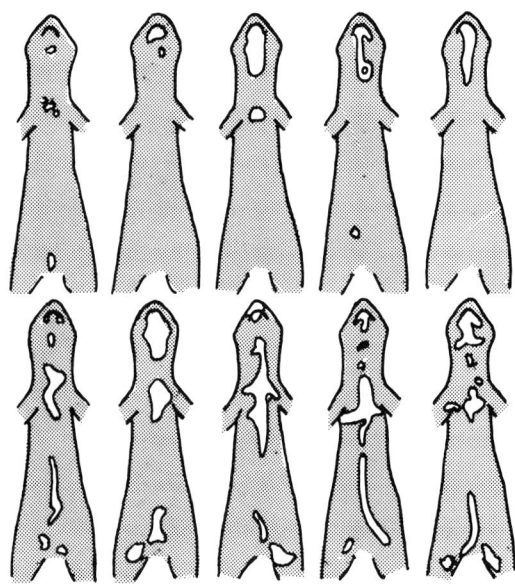

FIGURE 2.2 *The ventral spot patterns of a representative sample of trapped feral mink.*

The presence of extensive white hair on the upper lip of the European mink is said to be diagnostic, but care has to be taken in using this feature as a distinguishing characteristic since feral American mink can also possess it (see Chapter 3 for further information).

As mink grow older white hairs usually become apparent on the nape of the neck and around the mouth. These will have grown from scar tissue marking the sites of wounds obtained during fighting between males or during mating activity. Very old mink can acquire a quite grizzled appearance.

Mink moult twice a year. The summer coat may have a reddish tinge and is acquired in April. The dense winter coat grows out during September and October to reach 'prime' condition in late November to early December.

BIOMETRICS OF WILD POPULATIONS

Adult mink are medium-sized carnivores. Juvenile mink show highly variable weights during the summer months, but grow quickly to assume adult body size by about 10 months, in time for their first mating season. There is a great variation in growth rate within the sexes, but that of males is considerably in excess of that of females.

AMERICAN MINK

Males may continue to show an increase in body-weight after their first year, but this is rarely the case for females. The growth rates and survival prospects of males that fail to settle in a territory can be considerably less than those that do. This is indicative of the strong competition for resources in most populations.

Table 2.1 gives the measurements of feral American mink from various study locations in the UK and the USA. Adult male mink typically weigh about 1.2 kg whereas females are approximately half that size. Fairley (1980) captured one male feral Irish mink weighing 1.56 kg. Small, but mature, males of around 800 g are not uncommon.

TABLE 2.1 *Weights and measurements of mink.*

Location			Mean	Range
English river[a]			Weight (g)	
Male		Adult	1153	850–1805
		Juvenile	1009	685–1329
Female		Adult	619	450–810
		Juvenile	605	437–738
Idaho river[b]				
Male		Adult	780	
Female		Adult	525	
North Dakota river[c]				
Male		Adult	1523	
Female		Adult	853	
			Length (mm)	
Male		Head/body	397	330–450
		Tail	193	150 220
Female		Head/body	338	320–360
		Tail	168	135–190

Source: Chanin (1983)[a], Whitman (1981)[b], Eagle & Whitman (1984)[c].

Adult body-weight can change over the annual cycle, particularly in males. This is illustrated in Figure 2.3 with data from a coastal population of mink in Scotland. A similar pattern has been noted for populations of feral mink in south-west Britain, Ireland and Sweden, and for ranch mink (MacLennan & Bailey 1969). The weight of males increases during the autumn and peaks just before the mating period (rut) in January/February. This has an adaptive significance since the larger body size seems to confer an advantage during intrasexual fighting for access to females. Since weight declines after the mating season, males do not have to maintain this increased body size for the rest of the year. Weight changes in males are linked to the reproductive cycle and are probably related to the growth of the testes.

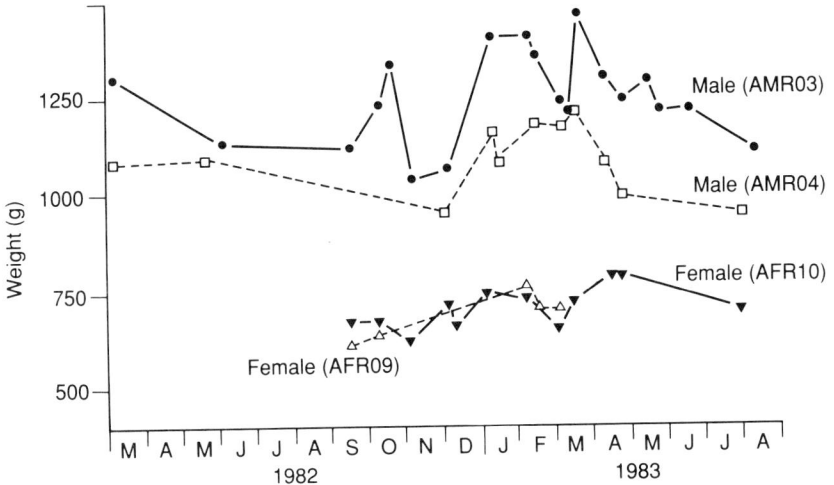

FIGURE 2.3 *Monthly changes in body-weight of male and female mink.*

EUROPEAN MINK

Both sexes are considerably smaller than American mink. Youngman (1982) noted the mean weight of seven males as 870.7 ± 152.7 g (range 600–1005 g). The weights of two females were 505 g and 580 g, respectively.

THE SKELETON

The general mustelid elongate body form with a long neck and relatively short stocky legs (see Figs 2.1 and 2.4) evolved as an adaptation enabling the pursuit of rodent prey into their burrows. This adaptation has only been achieved at considerable cost, since there will be increased metabolic demands for maintaining this inefficient body shape. This cost must be

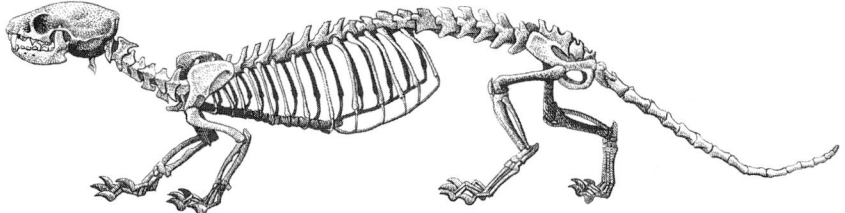

FIGURE 2.4 *Skeleton of the mink.*

compensated for by increased foraging efficiency, which should result from being able to enter narrow burrows in pursuit of prey.

THE SKULL

The brain size and the musculature of the jaw and neck are the main factors determining the shape of the mammalian skull. Wiig (1986) has demonstrated that for mink it is the requirement for biting power that has been the main determinant of skull shape (Fig. 2.5).

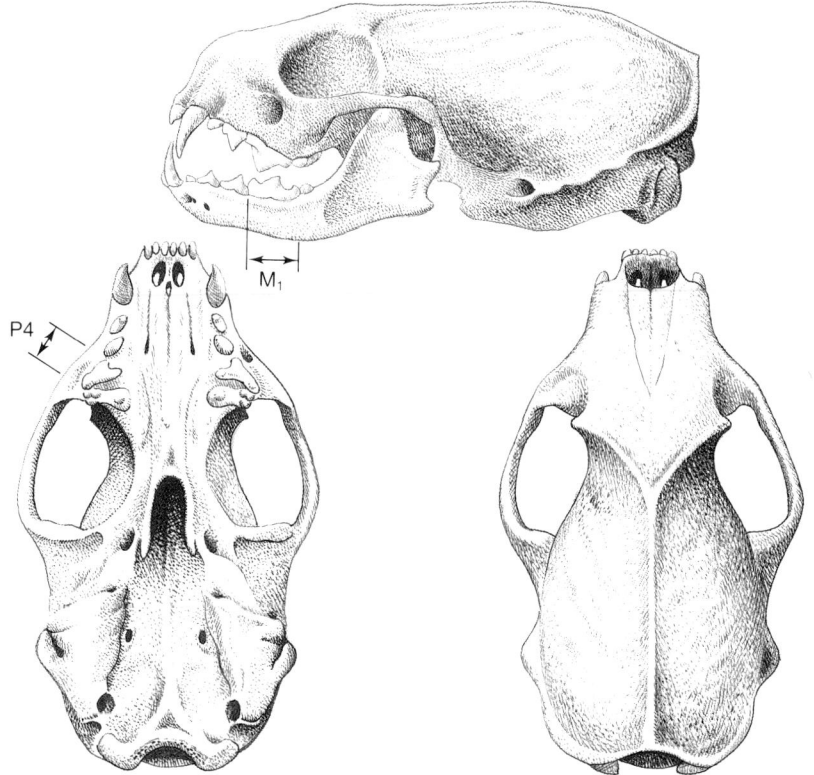

FIGURE 2.5 *Skull of mink.*

The skull of a male mink is some 15% larger than that of a female (Wiig 1982). There are also differences in the conformation of the skulls of the two sexes. Most of the sexual dimorphism in skull shape can be functionally related to the jaw mechanism. The larger male skulls will have longer jaws and larger jaw muscles with the potential for a larger absolute gape (Wiig 1986). A weak zygomatic arch (cheekbone) indicates that the main masticatory (chewing) function is powered by the temporalis muscle which originates from the brain-case roof. However, sexual dimorphism of the jaws and teeth is not as great as that of the whole head, and sexual dimorphism of the head is not as great as that of the whole body.

In most carnivores, it is the anterior part of the temporalis muscle and the zygomatico-mandibularis muscle which contribute to the main force when the jaw is open and the canines are in use (Fig. 2.6). The masseter and posterior fibres of the temporalis muscle are responsible for providing the shearing power of the carnassial teeth. The digastric muscle is usually large in fish-eating carnivores such as the otter. In mink it is not appreciably larger than in other terrestrial carnivores (Wiig 1986).

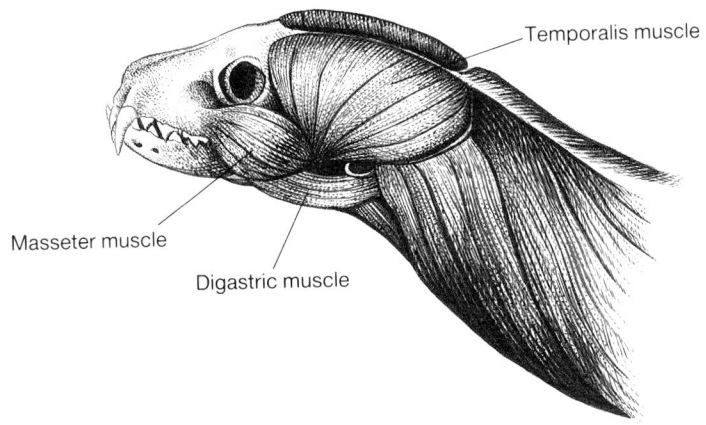

FIGURE 2.6 *Musculature of the mink head and neck.*

Dentition

The jaws are short and powerful containing small premolars, elongate canines and well-developed carnassials. The dentition of the mink, like that of most mammals, consists of two sets of teeth. The deciduous teeth do not erupt until the third week after birth. They remain for only a short time and are replaced by the permanent teeth in the second and third months of life.

The dental formula according to Aulerich & Swindler (1968) is:

$$\frac{I3}{3} \quad \frac{C1}{1} \quad \frac{P3}{3} \quad \frac{M1}{2}$$

where I, C, P and M refer to the incisors, canines, premolars and molars, respectively.

The canines are long, conical recurved teeth that resemble tusks or fangs. In the mink they are approximately 1–1.5 cm long and sharply pointed. The upper canines are longer than the lower pair. When the jaw closes the lower pair fit into a space between the upper canines and the incisors. These teeth puncture the prey, and are deployed in the killing bite which typically involves penetration of the back of the skull. They may also assist in holding struggling prey.

The last premolar of the upper jaw and the lower molar have become large and structurally complex. These two teeth are known as the carnassials and they intersect with a shearing action, important in shearing through tough flesh and bone. They have an action resembling a set of pruning shears in which the upper set overlaps the outside of the lower. Their effectiveness can be attested to by most mink researchers! I can graphically remember my early days as a research student working with so-called 'tame', bottle-fed and hand-reared mink babies. The object, my research supervisor informed, was to teach them an inhibited bite. This involved allowing the 'adorable' little mink kittens to freely bite my hand in the hope that they would 'learn' that there was no advantage to be gained in this behaviour. It actually seemed to work, at least as long as they were small. As they grew larger and more aggressive in their behaviour, teeth were frequently employed to investigate a proffered object, be it food or finger! As they matured into adults their familiarization with the scientist often meant that they took liberties, and while their bites were perhaps not as powerful as those directed towards prey, they were very painful and often turned septic. Most bites occurred when handling angry mink, and the very worst occurred if your hands were tainted with food. One of the worst I have sustained was in handling an almost 'moribund' animal for veterinary investigation. The animal suddenly made a recovery nothing short of miraculous, and savaged two fingers very badly. Often when a mink bites it will not immediately let go and the only way to get it to release it's grip in these circumstances is to provoke the animal into biting your other hand. The secret is to then get both hands quickly out of the way!

COMPARISION OF AMERICAN AND EUROPEAN MINK

Recently, feral American mink have spread extensively through northern Europe, including areas where the European mink is endemic. The two species are often confused in areas where they are sympatric. This is not surprising since, superficially at least, they are very similar. The American mink is on average 60% larger than its European counterpart, but there is considerable overlap since a female American mink is of roughly equivalent size to a male European mink. Size is therefore not a reliable criterion to use for identification.

In the field, close examination of the external features of trapped animals can often reliably separate the two species. Field guides (e.g. van den Brinck 1967, König 1973, Corbet 1980) often allege that the two species can be distinguished by the absence of white fur on the upper lip of the American mink (see Fig. 2.7 and Plate 9). Recently Linn & Birks (1989) have demonstrated that American mink in North America and feral mink in the British Isles can, quite commonly, possess white patches on the upper lip.

FIGURE 2.7 *European mink. Photo: Titi Maran.*

If there is a large regular white patch resembling a moustache on the upper lip, then the mink is likely to be *M. lutreola*, while a mink with none or just white flecks on the upper lip is probably *M. vison*. Therefore, while the absence of a 'moustache' is a fairly certain indicator of *M. vison*, the presence of white fur on the upper lip should be interpreted with caution.

The skeletons of the two species show considerable differences. American mink are reputed to have 19 vertebrae in their tails compared to 21 in the European mink (Pearson 1971), a feature that Youngman (1982) does not mention in his definitive study of the phylogenetic relationships of *M. lutreola*.

The skull of the mink is illustrated in Figure 2.5. The principal distinguishing characteristics between the skulls of the two species are the shape of the bullae: those of *M. lutreola* are narrow and almond shaped and overhang the surrounding bone, whereas those of *M. vison* are broad and triangular in palatal view. The shape of P^4 and M_1 also differ, the former is asymmetrical in *M. vison* with the lingual arm longer than the buccal arm. Similarly the lingual lobe of M_1 is relatively and actually longer in *M. vison*. The shape of the mandibles differs, the area of attachment of the digastric muscle is larger in *M. vison* and the rami of the mandibles are longer and thicker.

RELATIONSHIPS TO OTHER MUSTELID SPECIES

Given the evidence in this and the preceding chapter, it appears that the two species of mink are not closely related despite being grouped in the same genus. R. Wirth (pers. comm.) has suggested that perhaps the little-known Colombian and Amazonian weasels (*Mustela felipei* and *M. africana*, respectively) are close relatives of the American mink. Among the rarest of South American carnivores, these species have yet to be studied in detail. The fact that both have distributions associated with water perhaps suggests some affinity with *M. vison*. The Amazonian weasel is thought to have naked feet with interdigital webbing, and its reputed swimming capabilities suggest a semi-aquatic habit (Tate 1931).

CHAPTER 3

Finding Mink: Tracks and Signs

Only three forms of evidence reliably indicate the presence of mink: sightings, footprints and droppings (called scats). Partly eaten prey are all too commonly attributed to mink, but such accounts should be regarded as suspect unless accompanied by reliable evidence, such as tracks or droppings.

Mink tracks normally show five toes (see Fig. 3.1), although, occasionally, the impressions of only four toes may result when the mink is running at speed. The somewhat pointed, pear-shaped toes radiate forwards from a lobed central pad. Claw marks are often visible at the end of the pads, especially when prints are left in soft mud. The prints are 2.5–4.0 cm long and 2.0–4.0 cm wide. Those of adult male mink are somewhat larger than those of females, but great care has to be taken in distinguishing sex on the basis of print size, since during summer and autumn immature males (with female-sized feet) will be present in the population.

There is potential for confusion between the tracks of otters (*Lutra lutra* in Europe and *L. canadensis* in North America) and mink, since both species use the same habitat and occasionally the same dens. Generally otter prints are larger, particularly across the heel, and the toes tend to be more rounded in outline.

In Europe care also has to be taken not to confuse the prints of mink with those of the polecat (*Mustela putorius*) or the feral ferret (*Mustela furo*) in areas where there are overlapping populations. Where rainfall or flood has partly obliterated tracks, there is a possibility that prints of mink may be confused with those of stoat, water vole or squirrel. In North America there is the potential for confusion with field signs (tracks and scats) of the long-tailed weasel (*Mustela frenata*).

Tracks of mink are most commonly found within 2 m of the water's edge along streams and lake margins. The clearest tracks are left in fine silt and sand deposited after a flood. It is particularly fruitful to search for signs under bridges and overhanging vegetation since mink do not like to move over open ground. The quantity of tracks does not give any clear indication to the number of mink present since a single animal may traverse an area repeatedly during its nightly wanderings. In some areas mink may travel downstream

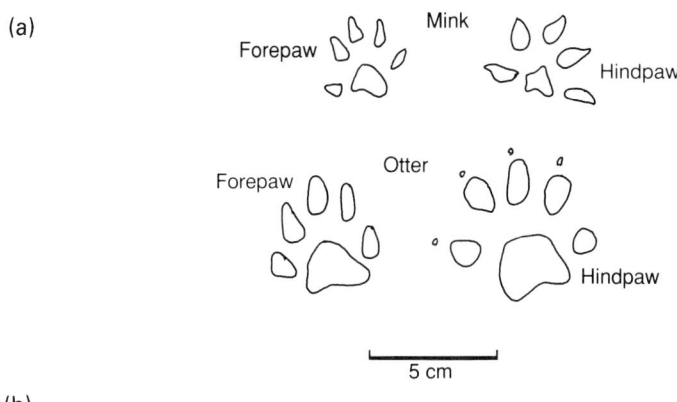

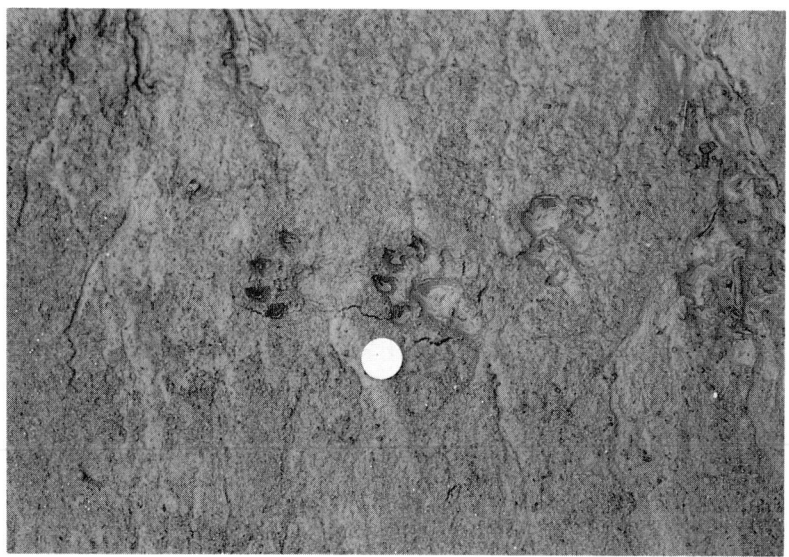

FIGURE 3.1 (a) Footprints of mink forepaw (left) hindpaw (right), and otter forepaw (left) hindpaw (right) and (b) mink footprints in mud.

using the water current and leave no sign, while there is some evidence that upstream travel may be more frequently undertaken overland. In coastal habitats, clear paths through vegetation along the top of the shore may be apparent.

Mink scats are cylindrical, usually between 5 and 8 cm long and less than 1 cm in diameter, often tapering towards the ends (Fig. 3.2). Fresh scats have an unpleasant foetid odour resulting from secretions of the proctodeal gland. During an initial survey for mink, droppings are rarely found, but once the sites of their deposition have been established they may be commonly located

FIGURE 3.2 *Mink scat with coin for size reference.*

in future. This is because the mink regularly deposits scats on prominent objects and features in its territory. Likely places to find mink scats are under bridges and on their abutments, prominent rocks, fallen tree trunks and at the confluence of rivers and streams.

Otter scats (referred to as spraints) are deposited in similar areas, but mink scats can be distinguished by their shape, which is usually more pointed, and their unpleasant smell. Faeces of polecats are very similar to mink. The colour and consistency of scats partly depends on the mink's diet and the scats' freshness. Aged scats dry up and are often olive-brown. Freshly deposited scats are dark, green-brown to black, often containing quite obvious fur and feather remains. Scats containing these items are commonly twisted. When mink have been feeding primarily on fish, the droppings are more granular and difficult to distinguish from those of otters, although the smell of mink scats is still relatively unpleasant compared to the 'newly-mown hay' aroma of spraint deposited by the otter! If mink have been feeding extensively on Crustacea (shore crabs or freshwater crayfish), then their droppings may be pale or even bright red due to chemical changes to the shell material that will have occurred in the mink's gut. Occasionally when a mink is disturbed it may only leave a greenish-yellow watery secretion.

During the breeding season female mink may often deposit large piles of scats outside their dens (see Plate 15). These can provide valuable information on the diet of an individual and her kits over a 1- to 2-month period. It is likely that mink may also deposit scats within their dens.

If mink are found in an area, animal carcasses or remains are often assumed to be the result of mink predation. Verification of kills attributable to mink is very difficult unless the killer is caught in the act. The 'bloodthirsty' image of mink has meant that they are often blamed, even when other predators are present in the area. Mink will occasionally take carrion, and hence, even the presence of mink tracks or scats at a kill does not necessarily implicate the mink as the primary cause of the incident.

The typical killing bite of the mustelid predator is directed towards the base of the skull (see Chapter 6), and usually results in small skin punctures approximately 9–11 mm apart. However, this method of dispatch is also used by other mustelids, and therefore cannot be relied upon to indicate mink as the culprit.

CHAPTER 4

Phylogeny, Distribution and Status

Mink are members of the family Mustelidae. One of the largest groups of the order Carnivora, the Mustelidae contains 67 species from 26 genera divided between the four subfamilies. From fossil evidence we know that mustelids arose from the arctoid carnivores. Since their origin, at the beginning of the Oligocene, the mustelids followed a separate evolutionary line, having deviated from the more closely interrelated dogs, bears and pandas. Considerable adaptive radiation of the mustelids occurred during the Cenozoic. Many of the lines were 'short-lived' and are not well represented in the fossil record, making interpretation of their phylogeny difficult.

The extant mustelid subfamilies are the Melinae (badgers), Mephitinae (skunks), Lutrinae (otters) and the Mustelinae which includes the mink.

Of the four, the subfamily Mustelinae shows the greatest diversity, comprising 33 species in 10 genera. It is an excellent example of mammalian adaptive radiation with species ranging from the arboreal martens (*Martes* sp.) to the semi-aquatic mink, with species inhabiting tropical regions (Amazonian weasel, *Mustela africana*) and arctic tundra (wolverine, *Gulo gulo*). They range in size from the diminutive least weasel (*Mustela nivalis rixosa*) weighing just 30–70 g, to the massively powerful wolverine which can reach 30 kg.

Mink are unusual in possessing semi-aquatic habits; they commonly dive underwater to obtain their prey. This represents a degree of convergent evolution with their more distant cousins, the otters (subfamily Lutrinae). As we shall see, the range of adaptations they possess which equip them for this semi-aquatic habit are not as well developed as in the otters. However, the mink is not a remnant from the otter's ancestral line, but a recent convergent, setting out on a similar evolutionary path. This has led to the suggestion that, in evolutionary terms, mink resemble a transitional state between the terrestrial mustelids and the otters. The otters are thought to have evolved some 40 million years ago, that is before the minks separated off from other mustelids in the Miocene (25 million years ago).

Various methods have been used to investigate the phylogeny of the subfamily Mustelinae. Traditional phylogeny is based primarily on the comparison of morphological features between contemporary specimens, and relating them to the morphology and distribution of presumed ancestors in the fossil record. However, the different selective pressures on each character over evolutionary time results in different rates of divergence, so the apparent relatedness of two species greatly depends on which characters one chooses to measure. Nevertheless, most animal classification has been established in this way. More recently, comparison of karyotypes (the number, size and structure of the chromosomes when viewed under the microscope) has been a widely accepted criterion for 'relatedness'. Even more informative are the modern techniques for assessing the likeness (homology) or divergence at the molecular level of certain proteins extracted from tissue or blood samples of the animal. One of the simplest and most widely used of these techniques employs antibodies raised in laboratory rabbits or mice. By injecting an extracted protein into a rabbit, an immune reaction results in the proliferation of antibodies which recognize and bind the foreign protein in a highly selective manner. Equivalent proteins from a related species will usually be recognized by the antibody, but it will bind less strongly according to the degree of divergence of the proteins. Such techniques employing antibodies are referred to as 'immunology'. Techniques of DNA 'fingerprinting' are now widely applied to sort out taxonomic tangles. It is hoped that this suite of techniques will soon be applied to the mink.

Comparative studies of the impoverished fossil record and the body form of living species suggest that modern day mustelines with their long bodies and short stocky limbs have changed little from their primitive forest-dwelling ancestors. Examination of diverse features, including the morphology of the skull, the number and structure of the chromosomes and immunological evidence, indicates four main groupings within the subfamily Mustelinae:

(i) the *putorius* group including the European polecat (*Mustela putorius*), the Steppe polecat (*M. eversmanni*) and the black-footed ferret (*M. nigripes*);
(ii) the *lutreola* group including the Siberian weasel (*Mustela sibirica*) and the European mink (*M. lutreola*);
(iii) the *mustela* group comprises the long-tailed weasel *Mustela frenata*, the mountain weasel (*M. altaica*), the stoat (*M. erminea*) and the weasel (*M. nivalis*);
(iv) the American mink (*M. vison*) was placed in a group of its own, the *vison* group. *Mustela africana* has not yet been adequately described, but may be a candidate for the *vison* group.

The inclusion of the European mink with *Mustela sibirica* reflects the overall similarity between their skulls and the identical nature of their karyotypes. They each possess 38 chromosomes of similar size and morphology.

What is important, is that the American mink and the European mink are *not* one and the same species, although they are often included in the same subgenus *Lutreola*. Indeed, from these analyses it would appear that they are not as closely related as might be expected. Although they share a semi-aquatic niche and some similarities of the skull, the dissimilarities are fundamental (Youngman 1982). In some aspects of its cranial morphology, *M. vison* resembles the polecat, and in others the martens. On immunological evidence the American mink is more closely related to the sable (*Martes zibellina*) than to the European mink. All the evidence suggests that the American mink is neither an aquatic polecat nor a near relative of the European mink. This is contrary to early thoughts about the interrelationships within the genus *Mustela*.

It was Kurtén (1968) who made the original suggestion that the European mink arose from a range extension of *M. vison* across the Bering Land Bridge which joined Siberia and Alaska during the Ice Age. As the ice retreated, the two species became isolated and evolved the minor differences we observe between modern day specimens. Further evidence to support the hypothesis that the European mink is a relative newcomer to the fauna of Europe comes from the paucity of fossil mink remains that have been found in Europe.

Corbet (1966) implied a phylogenetic relationship between *M. vison* and *M. lutreola*, with the Siberian weasel (*M. sibirica*) as a connecting intermedi-

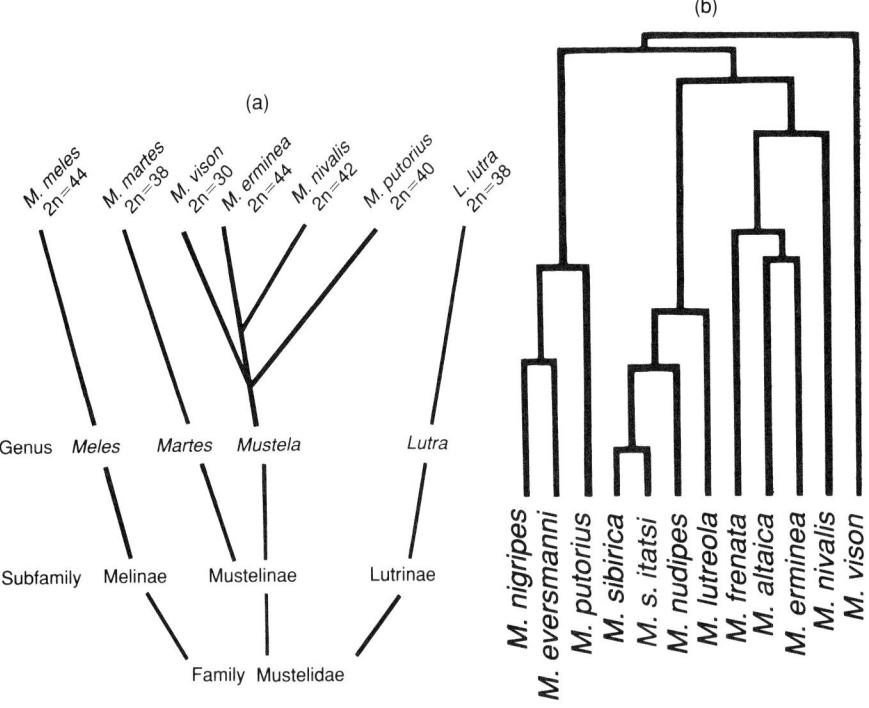

FIGURE 4.1 *Possible phylogenetic relationships of mustelids based on (a) chromosome numbers, (b) morphological characteristics. Source: (a) Brinck et al. (1986); (b) Youngman (1982).*

ate species. This implies that the American mink is more closely related to the Siberian weasel than the European mink, but so far this claim has not been substantiated. The Siberian weasel is also variously known as the kolinsky or Chinese weasel, and its Japanese subspecies, *M. siberica itatsi*, is known as the Japanese mink. In comparison with the kolinsky, pelts of Japanese mink show a closer resemblance to the European mink, and also possesses the characteristic white markings on the upper lip which resemble a moustache (Linn & Birks 1989). Youngman (1982) also emphasizes the closeness of the *M. sibirica–M.lutreola* lineage, although there seems to be no natural hybridization between the two species.

It has been suggested that the Siberian weasel is competitively superior, and in the eastern Palaearctic region excludes the European mink. Currently little is known of the behaviour and ecology of any of these species of Asiatic mustelids. No doubt when such knowledge is forthcoming the classification will be revised and our knowledge of the evolutionary origin of *M. lutreola* clarified.

Tree diagrams depicting these phylogenetic relationships are shown in Fig. 4.1(a) and (b).

DISTRIBUTION OF THE VARIOUS SUBSPECIES OF MINK IN NORTH AMERICA

The species was first described by Schreber in 1777. Since then 15 distinct subspecies of *M. vison* have been recognized across the North American continent (Fig. 4.2). The mink bred on fur farms probably represent the interbreeding of three to six of these subspecies, and it is from these that European and Scandinavian populations of farmed, and subsequently feral, American mink have been derived. The differences between the various subspecies of mink are subtle and involve quantitative traits such as size, and qualitative features such as coat colour and quality. It should be noted that these subspecies were described on the basis of relatively 'outdated' taxonomic techniques. It has yet to be determined whether the existing classification would stand up to the rigours of DNA 'fingerprinting'. *M. v. vison* ranges through eastern Canada to northern Pennsylvania, excluding the coast south of New Brunswick. Another subspecies, *M. v. lowii*, is found in northern Quebec and Labrador, while *M. v. letifera* ranges from northern Wisconsin and northern South Dakota south to northern Illinois and Missouri. *Mustela v. lacustris* ranges through the interior of Canada from the Great Bear Lake and from the western shores of Hudson Bay south through Alberta, Saskatchewan and Manitoba provinces to southern North Dakota. The large 'Yukon' or 'Alaskan' mink is of the subspecies *M. v. ingens*. Its range is adjacent to that of *M. v. melampeplus* at the base of the Alaskan peninsula and *M. v. lacustris* in the Mackenzie valley. *M. v. mink* is an 'eastern' variety ranging over the coast of New England south to North

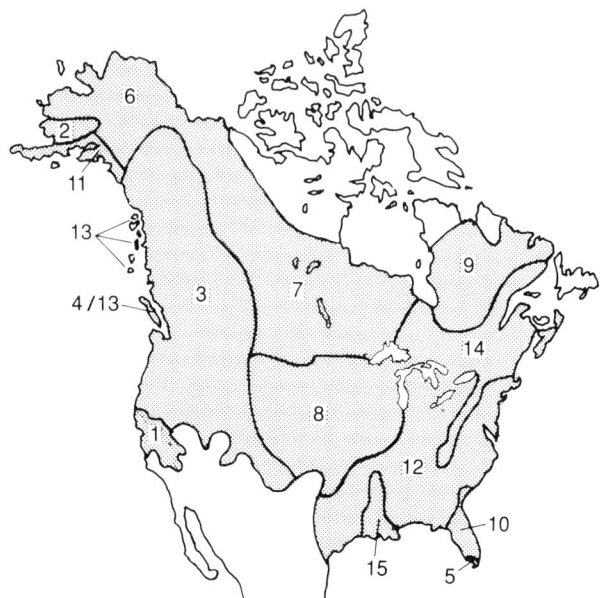

FIGURE 4.2 *Distribution of the various subspecies of mink in North America. Source: Hall (1982). 1*, Mustela vison aestuarina; *2*, M. v. aniakensis; *3*, M. v. enurgumenos; *4*, M. v. evagor; *5*, M. v. evergladensis; *6*, M. v. ingens; *7*, M. v. lacustris; *8*, M. v. letifera; *9*, M. v. lowii; *10*, M. v. lutensis; *11*, M. v. melampeplus; *12*, M. v. mink; *13*, M. v. nesolestes; *14*, M. v. vison; *15*, M. m. vulvivaga.

Carolina. There is another supposed subspecies unique to Florida, the Everglades mink, *M. v. evergladensis*, which was listed as threatened by the Florida Game and Freshwater Fish Commission in 1980. To the west, *M. v. energumenos* predominates, with *M. v. aestuarina* in California.

THE ORIGIN AND SPREAD OF THE FERAL POPULATIONS

Since the late 19th century mink have been bred on fur farms in North America. The original farmed mink are thought to have been mainly derived from *M. v. vison*, *M. v. melampeplus* and *M. v. ingens*. The first American mink were brought to Europe in the 1920s for commercial farming, and the mink in Europe today are descendants of escapees from these fur farms. In addition to inadequate housing, storm damage and vandalism will have contributed escapees to the feral population. In some countries, for example Russia, feral populations result from the deliberate releases of breeding stock into the wild during 1930–1950, to provide a natural crop. Breeding populations in the wild which are derived from released or escaped domestic animals are termed feral.

DISTRIBUTION AND STATUS OF FERAL AMERICAN MINK

Not surprisingly it is those countries that carried out commercial mink farming which now have populations of feral American mink. These include most European countries, particularly Britain, Ireland, Germany, Spain and France. In Scandinavia mink are widespread, and they are also found in Iceland. The Russian mink population, mainly derived from deliberate releases, has spread into Poland.

Some breeding of mink has also been carried out in South America since the 1930s, and there feral populations exist in Argentina and probably Chile (Olrog & Lucero 1981, R. Medel pers. comm.).

ENGLAND AND WALES

Mink have been farmed in Britain since 1929 and, given their superlative ability to escape confinement, some will probably have been free since that time. Certainly mink have been killed in the wild since the 1930s, although there was no confirmation of breeding in the wild until July 1956 when a female and kits were sighted near an old mill leat on the River Teign in south Devon (Linn & Stevenson 1980). Apparently the Ministry of Agriculture was aware of a colony of mink in the Singleton area of Lancashire before this time (Thompson 1962, Clark 1970). During the late 1950s and early 1960s other populations became established on the Rivers Avon, Test and Itchen in the counties of Hampshire and Wiltshire; the Avon in Gloucestershire, and the Rivers Wharfe and Lune in Lancashire and Yorkshire. In Wales mink established themselves on the River Teifi in mid Wales and the East Cleddau and Taf in Pembrokeshire.

Scientific acknowledgement of these populations came quite late, when Ian Linn of Exeter University presented a report of mink breeding in the wild to a meeting of the Mammal Society in November 1957. Following this publicity mink regularly became a media event. In an attempt to prevent further escapes, fur farms were required to provide a minimum standard of security. The Ministry of Agriculture imposed legislative control in 1962 with the Mink (Importation and Keeping) Order which was made under the Destructive Imported Animals Act of 1932. The regulations require farms to be licensed, a condition of issue of the licence being that the enclosure meets a specified standard of security. Licence holders are required to report escapes from their premises and occupiers of land have an obligation to report the presence of feral mink on their land.

The Ministry of Agriculture were initially sceptical of the mink's potential as a pest species in England, even though good knowledge of their biology was available from Scandinavia and Iceland. By 1962 this complacent attitude had changed and trappers employed by the Ministry accounted for over 5000 animals in England and Wales alone. This was a case of 'too little, too late', and by the 1970s mink had been recorded in every county. By 1967 the

Ministry had decreased its active participation in trapping to the extent of only operating a trap loan scheme.

The spread of mink on the River Teign in the south-west of England is well documented (see Fig. 4.3). The initial stages were slow, mink taking some 8–10 years to colonize a short stretch of the wooded valley. Then, within the course of 3 years, the mink population expanded explosively to cover an area some 20 miles by 50 miles. A similar apparent expansion of range was noted in the mid 1960s for populations in Lancashire (Clarke 1970). Here captures doubled in 1965, with the Lancashire population extending north to the Lune and south and west up the River Ribble into Yorkshire. At this time mink also appeared on other Yorkshire rivers including the Ure, Nidd and Wharfe. To a certain extent this increase in population was due to a greater awareness of mink and an increased trapping effort by Ministry officials. It seems likely that mink are present on many rivers for some time before their presence is realized; hence, any distribution map reflects the activity of recorders as much

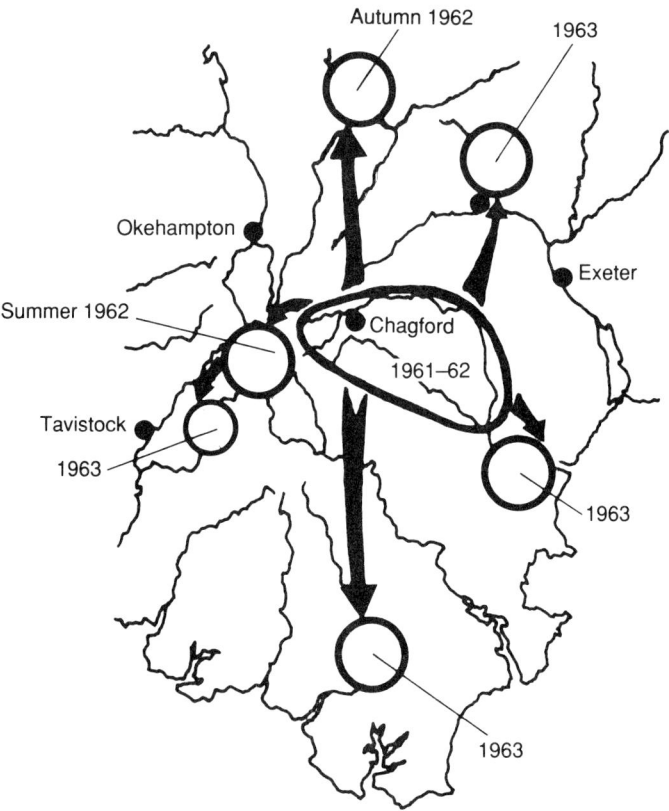

FIGURE 4.3 *The spread of a feral mink population in Devon, UK during 1956–1963. Source: Linn & Stevenson (1980).*

as the presence of mink. The current distribution map for feral mink in the British Isles (Fig. 4.4) has been compiled from a number of sources, the most significant being the Otter Survey of the British Isles funded by the Nature Conservancy Council and Vincent Wildlife Trust. During this systematic survey for the rare otter, the presence of mink was also recorded by trained field workers. Mink appear to be still expanding their range in parts of northern England and mid Wales. In other areas, the south-west England, Yorkshire and Lancashire, mink appear to have established stable populations.

FIGURE 4.4 *Distribution map for feral mink in the British Isles.*

SCOTLAND

John Cuthbert of the Department of Agriculture and Fisheries for Scotland has detailed the origin and distribution of Scottish populations of feral mink (Fig. 4.4). (Cuthbert 1973). The first farm was established somewhat later than in England, in 1938 in southern Scotland. By the late 1940s and 1950s there were about 100 farms. The industry subsequently went into decline and consolidation, until by 1971 only 29 working farms remained. But the numbers of mink kept dropped relatively little from 28 000 in 1962 to 23 000 in 1971 indicating that it was the smaller farms that closed down, while others expanded.

Although there had been a number of captures of mink dating from as early as 1938, true breeding populations were not located until 1962, on the River Ugie in north-east Aberdeenshire and the River Deveron in Banffshire (Hewson 1971). In 1963 evidence of breeding was obtained when kits were trapped on the River Urr near Kirkcudbright, southern Scotland. The appointment of four Ministry trappers in 1964 accounted for 324 mink being killed that year. Official participation in trapping was withdrawn in 1970.

Again it is apparent that the mink were present on many rivers for some time before their presence was reported. Mink were first reported on the Isle of Lewis in 1969, but must have been present from 1961 when the last mink farm closed down. Mink have been known on the Isle of Arran since the late 1960s (Gibson 1969). The population is thought to have arisen from an unlicensed farm rather than by animals colonizing it from the mainland, a journey of four and a half miles. Mink have been observed to swim between Arran and nearby Holy Island, a distance of half a mile. There were two farms on Shetland, but no problems of feral mink have been reported. Mink are known to have escaped there, but fortunately all were non-pregnant females. Good populations of mink are known on the Hebridean islands of Lewis and Harris, situated off the west coast of Scotland. In recent years permission has been sought to open a mink farm on the island of Westray in north-west Orkney, home for one-quarter of a million breeding sea-birds, but after vociferous debate, this was denied.

IRELAND

The first mink farms in Ireland were established in 1950–1953, and by 1960 there were 40 mostly small farms operational. The licensing of farms became mandatory in 1965. Deane & O'Gorman (1969) document various sightings and deaths of escaped mink, the first sighting occurred in 1961 from a farm in Omagh, County Tyrone. Mink then became established on the Shrule, Mourne and Glenelly river systems, and have been subsequently recorded in at least 16 counties (see Fig. 4.4), with the possible exception of Galway. The north of Ireland, particularly County Donegal, the midlands and the eastern coastal region contain the largest mink populations. In western and southern counties mink distribution remains restricted.

WESTERN EUROPE

Elsewhere in Europe the distribution of feral American mink is not well documented (see Fig. 4.5), although populations are known to exist at least in Britanny—France (Lafontaine 1988), Spain, Netherlands and the Schleswig-Holstein region of Germany (Heidemann 1983). In the latter case sightings and killings have been reported since 1950. Undoubtedly the species is far more widely distributed than this.

SCANDINAVIA

Sweden
Sweden gained an early entry into the fur industry, with the first farm started in 1925. Free-living, although not necessarily truly feral, animals were

FIGURE 4.5 *Distribution map for feral mink in Europe and Asia. Source: based on Lever (1985).*

reported in 1928 (Gerell 1967a). Following two mass escapes, at Bjurholm and Hjärtum, populations built up slowly in the 1930s around the eutrophic (biologically rich) lake habitats and coastal regions of southern Sweden. There seems to have been a low tendency for dispersal at that time probably because the population density was low and the quality of habitat so good that its carrying capacity took some time to achieve. The distribution of wild mink corresponded with areas where the number of mink farms was greatest. Subsequently, the increase in the feral mink population resulted in a greater trapping effort, which tended to distort the interpretation of their rate of spread. During 1949–1959 populations continued to increase, and the mink extended its range considerably. As a result of increased hunting pressure, a decline was noted in about 1961, but populations subsequently recovered. The annual catch of feral mink in Sweden was approximately 17 000 in 1963–1964. The distribution shows only mountainous regions in the north to be mink-free when the country was surveyed in 1967.

Norway
Here also, the distribution of feral mink shows a high correlation with the development of mink farming. The first farm was established in 1927, but the industry has grown considerably since the war to a maximum production of three million pelts in 1969. This was followed by a decline to a present day production of one million. By 1948 mink were well established over much of

southern Norway and by 1949 the first were captured north of the Trondheimsfjord. Farming was widespread and the numerous epicentres of escapes led to a thorough colonization of the country during the 1950s and 1960s, since when their distribution has changed very little. Only certain islands, usually those more than 5 km offshore, and the greater part of Finnmark province remain mink-free (Bevanger & Ålbu 1986a). Norwegian populations of feral mink are generally distributed along the coast. Low lying archipelagos with rich vegetation are particularly favoured.

Denmark
Approximately one million mink are kept in captivity, and there are reports of up to 300 feral animals being killed annually. Mink farming began on the Faeroe islands during the 1940s, but no reports of feral animals are available. It would be surprising if there are none.

ICELAND
Mink were first imported to Iceland for fur farming in 1931, and in 1937 the first record of a breeding den was recorded near Reykjavik. The population had spread throughout the southern and eastern parts of the country by 1960s and continues to spread around the coast (Skirnisson 1980).

In Iceland mink are extensively hunted in an attempt to keep the population low. An annual catch of 2000–2500 has been made since the late 1950s. A bounty scheme is in operation which may explain the relatively high annual catch rate.

EASTERN EUROPE, FINLAND AND RUSSIA
Feral American mink have been killed in Poland since 1962. These are thought to have originated from farms in the Wielkopolska-Kujawy Lowland and Msurin Lake Districts. By contrast, those killed in Bialowieza Forest may be derived from a population of 895 animals deliberately released in Byelorussian SSR during 1950–1953 (Ruprecht *et al.* 1983).

The first American mink were brought to the USSR for farming in 1928. In 1933 a shipment was made to Karelia for release into the wild. One hundred mink which were released in 1956 on the Kara-Ungur and Kugart Rivers in Kirigizia had shown a sufficient population increase to justify commercial trapping by 1961. Feral mink populations have also been established in western Siberia. Between 1936 and 1961 over 7000 mink were released in the Soviet Far East (Khabarovsk and Maritime territories).

In total, over 16 000 mink have been released in the USSR since 1962 (Lever 1985). In general they have fared best in Siberia and the Far East, especially in Primor'ye, Priamur'ye and in the Altai.

EUROPEAN MINK (MUSTELA LUTREOLA)

Considerable care has to be taken to verify that records are indeed of the native European mink and not of the American mink, since the two species are superficially similar and are now sympatric in parts of their range. However, there is sufficient verified information to state that the European mink is in decline, and may be considered a threatened species. The distribution map for the species is shown in Fig. 4.6.

The first author to mention the European mink was Agricola (1549):

> The mink, which also lives in woods, is the same size as the pine marten, but its hair is smooth and short and almost like that of an otter in colour; the mink's pelt, however, is far superior to the otter's, and is still preferable even if white hairs are mixed with the rest. The animal is still found in the dense, uninhabited woods which lie between the Oder and the Vistula.

But it was not until 1761 that the species was first described by Linnaeus, originally as *Viverra lutreola*, subsequently to be amended to *Mustela lutreola*.

The distribution and systematics of the European mink have been extensively studied by Youngman (1982) of the National Museum of Natural Sciences, Ontario, Canada using museum specimens and an exhaustive literature survey. He considers its historic range to have been from the Pechora river basin in the USSR to the Basque region of northern Spain. The species has never occurred in the British Isles, Scandinavia or Belgium. Heptner & Naumov (1974) list five subspecies of mink from the USSR and surrounding countries (*M. lutreola lutreola*, *M.l. novikovi*, *M.l. turovi*, *M.l. cyclipena* and *M.l. transsylvania*).

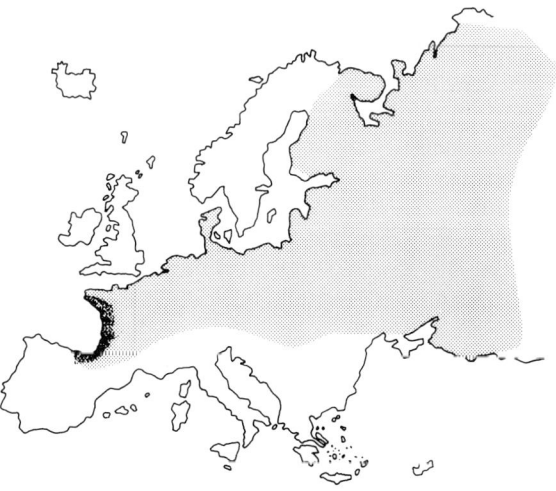

FIGURE 4.6 *The historic distribution of the European mink (stippled). Recent colonizations are indicated by shading. Source: based on Youngman (1982).*

The considerable Russian literature on this species has been summarized by Youngman (1982) and Schreiber et al. (1989). The USSR is the major stronghold of the species, but even here it is in decline with an estimated population of only 40 000–45 000 remaining. Maran (1990) considers that the range has declined by as much as 80%, and that the species may become extinct within the next 15–20 years. The watersheds of the Rivers Wasuwa, Ugra, Sosh, Oster, Chmara, Wolga and Dwina, in an area to the north-west of Moscow are thought to contain about half the European mink population in the USSR. Densities here are approximately 2.5 per 1000 ha compared to only 0.1 per 1000 ha west of the Ural mountains. The species is thought to be extinct in western Siberia, northern Kazakhstan and the republics of Moldowian, Baschkirsic and Tartaric. In the course of studies conducted at the Biological Institute of the Siberian Department of the USSR in Novosibirsk, Ternovsky released 338 captive bred European mink on the Kunaschir and Urup islands during 1981–1986 and 106 in Tadzickistan, even though the former area is outside of its normal range. The introduction of European mink to an area which the American mink cannot colonize naturally may be a way of safeguarding the survival of M. lutreola.

A number of nature parks and reserves in the USSR are likely to contain European mink. These include Lahemaa National Park in Estonia, and the reserves of Tsentralno-Lesnoi, Karpatskii, Dunaiskie Plavni and Kaneevskii. Other protected areas known to contain the species are Cernomora, and Ritsa-Avakhar and Adzhametsky in the Georgian SSR.

In the last century populations of the European mink have evidently suffered a serious decline. Competition from feral American mink is commonly deemed responsible, but habitat destruction is likely to be the main reason for the decline of the European mink in Russia. By analysing the numbers of pelts sold in the Leningrad region from 1913 to 1965, Danilov & Tumanov (1976) suggest that there may be an 11-year cycle in mink numbers. Since the introduction of the American mink, peaks in these cycles have not been as high. European mink represented 60% of pelts in the Leningrad region in the 1960s, but this had reduced to 40–50% by the 1970s. It is thought that where the European mink is over-harvested it is replaced by M. vison. Danilov & Tumanov (1976) propose a total hunting ban on all mink and a discontinued release of the American mink into regions predominantly occupied by the native species. Apparently a captive breeding plan has been implemented with the aim of re-introducing animals into the wild

FINLAND

The European mink decreased in numbers during the 1930s and was almost extinct by 1946. Numbers do not seem to have recovered, possibly due to the increasing presence of feral American mink in the 1950s. This is not to suggest the M. vison caused the decline, since the reduction in native mink numbers preceded the establishment of the feral population. The last records of the native species, for which museum specimens are available to verify the identification, are from 1965, and although there have been more recent reports (in 1981) these have yet to be substantiated and may be of American mink.

EASTERN EUROPE

The European mink has been reported from the Transylvania and Moldavia regions of Romania. An annual harvest of at least 2000 skins has been reported by Heer (1965). Youngman (1982) provides additional evidence for a large population in the Danube Delta.

Formerly the species was widely distributed in Poland, but Romanowski (1990) considers that it was a rare animal even by the 17th century. Historically *M. lutreola* occurred in two areas in Poland, the Odra river basin in the north where they became extinct in the 1830s and the Mazurian lakes and areas along the eastern border where they were probably found up until World War II. The most recent (authenticated) specimens originate from near Schwentainen in 1909, Elblag in 1915 and northern Poland in 1926. Although the European mink is now generally considered to be extinct in Poland, it is quite possible that it survives in small numbers, especially since it still occurs in neighbouring Byelorussia (Tumanov & Zverev 1986).

In Czechoslovakia recent unsubstantiated reports are thought to be of American mink. The last documented specimen of *M. lutreola* dates from 1888. In 1952 a possible specimen was taken from a brook near Tihany, Lake Balaton in Hungary. Drainage of marshland and wetlands for agricultural purposes has probably led to a decline elsewhere in the country.

There are no recent records of native mink for Switzerland, Austria or Eastern Germany, although historically it was probably found in all three countries. The European mink was common in Germany: Gesner (1620) reported thousands of furs on sale in Frankfurt. This flourishing fur trade was evidently too much for the mink, and it was thought to have become extinct by the middle of the 19th century (Youngman 1982). Nevertheless, Schreiber *et al.* (1989) considers the last authenticated record of European mink from Western Germany to be as recent as 1949 from the Aller River near Wolfsburg in Lower Saxony.

FRANCE

Mustela lutreola ranged throughout non-Mediterranean France from Brittany in the north to the Spanish border. Youngman (1982) postulates that the European mink was a late-comer to France. The 17th century was a period of severe temperature, and the spread of the mink into France seems to have coincided with an amelioration of the climate. He considers the spread to have encompassed the Paris Basin and the Departments of the Massif Central as far south as Drome in the Rhone valley. According to Van Bree & Saint-Girons (1966) the mink had disappeared from the Paris Basin by the early 20th century.

Now only two regions, Britanny and south-western France are reported to contain significant populations. Recent records show Britanny, the northeast, central and southern parts of the Department of Morhiban to hold good populations. In some parts of its former range (Côtes-du-Nord), numbers seem to have declined since 1970. In the south-west the species is said to be 'quite common' on the Rivers Beau, Né, Charente and Caran and in the valleys of the Soute and Seugne.

SPAIN

The European mink was first reported in Spain in 1951, when specimens were captured in the vicinity of Tolosa in north-central Spain. European mink have only been reported from a comparatively small area in the north, ranging from Navarra in the east to Asturias in the west. Youngman (1982) suggests that the species has spread into this region from France as recently as the 1940s.

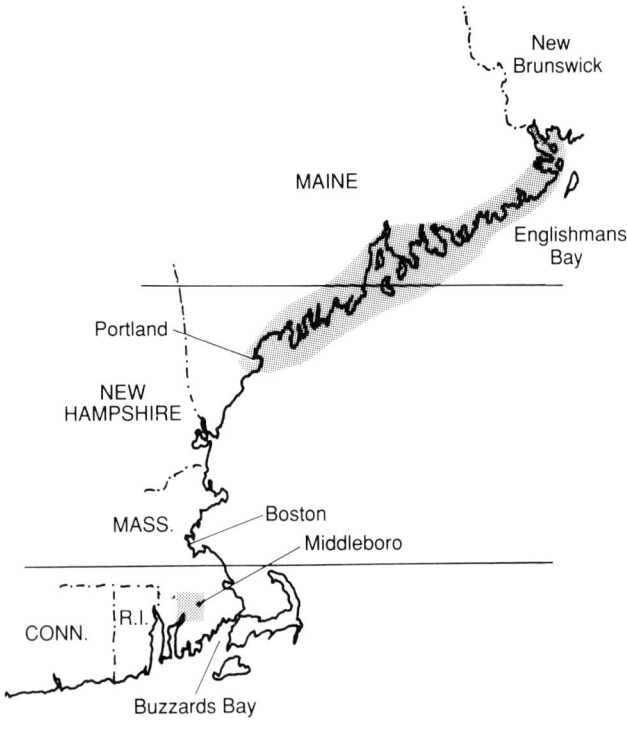

FIGURE 4.7 *The known range of sea mink,* Mustela macrodon. Source: Waters & Ray (1961).

THE SEA MINK, MUSTELA MACRODON

For one species of mink (variously referred to as the sea mink, saltwater mink or big mink) over-exploitation for fur led to extinction. It was first described from fragments of its skull found in shell heaps at Brooklin, Maine. It appears to have been a larger animal than *M. vison*: the dimensions of the one mounted specimen available (taken from the Bay of Fundy in 1874) suggest that the body was almost 1 m long plus a 25 cm tail (Goodwin 1935). Its range (as determined from archaeological finds) is from New Brunswick, Maine, New Hampshire to Massachusetts (see Fig. 4.7). The most recent finds of *M. macrodon* bones were made in 1961 and 1962 by Waters and colleagues in Massachusetts (Waters & Ray 1961, Waters & Mack 1962) at the site of an archaic Indian village. But the species is best known from a skin of an animal ('the last?') which was taken from the Bay of Fundy, and is believed to date from 1874 (Goodwin 1935). If this is the case the species may have been extinct for over 100 years. The site of the more recent finds of bone (which are of unknown age) was 12 miles from the nearest sea water, and it is suggested that it might have been brought there by the Indian inhabitants when they returned from their summer fishing expedition. The burnt bones were found in the equivalent of a refuse pit. The 1962 find was also some distance from sea water at a ceremonial burial site. The scientists suggest that the sea mink was one of the many species of animals consumed during ceremonial feasts. Little else is known about this species apart from its seashore living habit where it presumably fed upon marine organisms.

CHAPTER 5

Adaptations to Habit and Habitat

THE PELAGE

Many species of mustelid have been exploited throughout their range on account of their luxurious coats which are highly prized by the fur trade. With an annual harvest of 300 000–400 000 wild mink in North America alone, and perhaps as many as 25 million ranch-bred animals world-wide the economic importance of this species is self-evident.

Like the fur of many northern distributed species of mammal, the mink's pelage is dense being made up of two types of hair, guard hairs and underfur. The longer, stiffer guard hairs project beyond the underfur. The underfur consists of one type, while the guard hairs can be classified into three groups on the basis of their length. The upper part of the guard hair is lanceolate and rather thicker in diameter. The lower part or shaft is considerably narrower. Guard hairs are associated with arrector pili muscles which can cause the hair 'to stand on end' when the mink is 'excited'. The guard hairs have a mean density of 780 hairs cm^{-2} in the mid back region. Each shaft is surrounded by underfur which is half to two-thirds of the length. In the winter coat there will be from 25 to 40 underfur hairs associated with each guard hair, whereas in the thinner summer coat this will have reduced to five to eight hairs (Joergensen 1985). Each guard hair with its complement of underfur fibres, opens through a common orifice in the skin (see Fig. 5.1). The underfur produces a dense matted felt-like layer which provides an efficient insulating layer by trapping air next to the skin providing protection against the low northern temperatures. In semi-aquatic species such as the mink, there is the additional problem of immersion in cold water to cope with. The same fur that insulates against the cold also provides a high degree of water repellancy. A violent shake of the body when the animal leaves the water is all that is required to fluff up the coat allowing it to regain its insulative property. The coat of the mink is considerably more dense than that of the polecat or its commercially farmed relative the ferret or 'fitch'.

It is the guard hairs which give the colour to the coat. Each hair consists of three concentric layers, the cuticle, cortex and the medulla. Melanin pigment

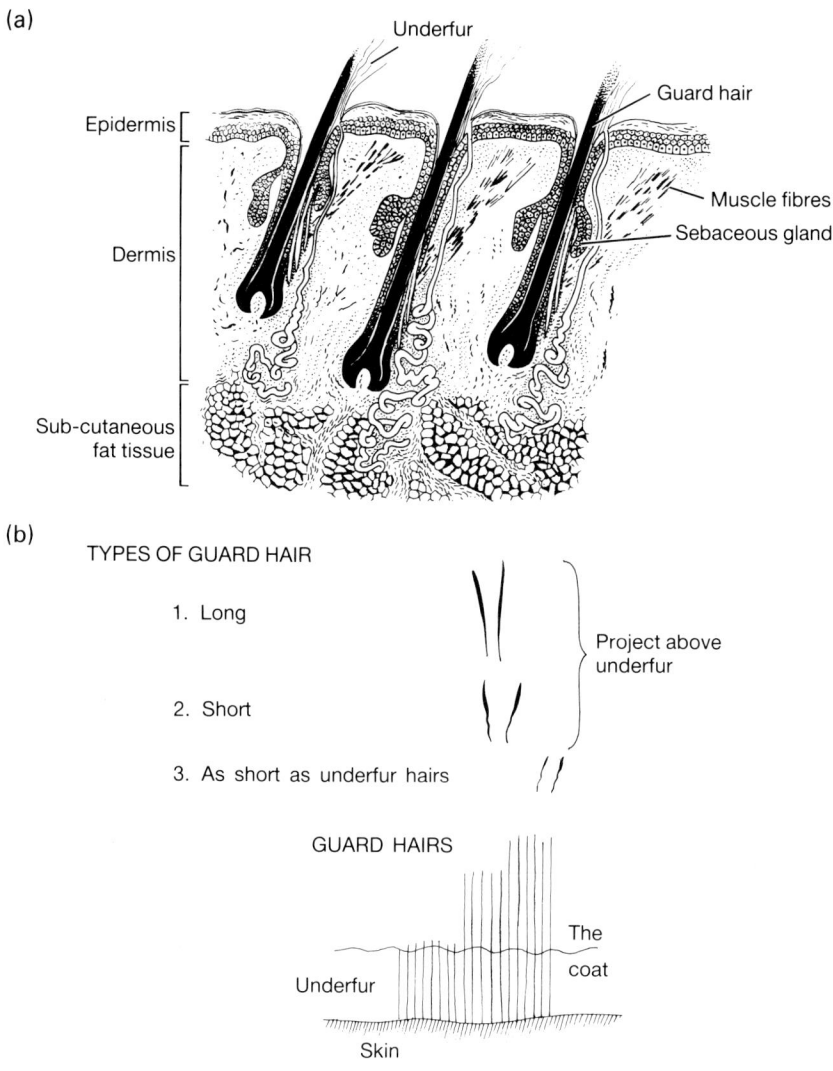

FIGURE 5.1 *(a) The hair follicle. (b) The pelt consists of long and short guard hairs, which are lanceolate in shape, and a dense underfur. Source: Joergensen (1985).*

granules are contained in the cortex and medullary layers of the guard hair. In addition to the dark brown 'wild type' coloration there is a plethora of colour varieties of mink that have been bred selectively for the fur trade (see Chapter 10).

Many mammals exhibit seasonal changes in their pelage. In the mink, like many other species of mammal, these have been shown to be photoperiodically controlled. Fur growth is under the control of daylength. Decreased daylength as winter approaches initiates a change to a heavy dense coat and an increase in daylength with the spring leads to a sleek, less dense summer coat. The mink sheds its coat twice a year. Growth of the summer coat in the adult mink begins in mid April and is completed by mid July to early August when the autumn moult begins. Growth of the dense winter coat takes place during September to early November. Studies conducted by Rust *et al.* (1965) have demonstrated that hormones from the pituitary gland located in the hind brain are necessary for the normal pelage cycles to occur. In Britain feral mink also moult twice each year, during spring and autumn (Hewson 1971, Linn & Stevenson 1980, Chanin 1983).

LOCOMOTION

The biomechanical demands of swimming through water and terrestrial locomotion are vastly dissimilar. To cope with these demands divergent trends have evolved across a wide range of amphibious and aquatic mammals in the degree of specialization of their appendages. The webbed appendages characteristic of many aquatic species enhance the surface area available for thrust generation during swimming. This is usually more pronounced in the hind limbs. The mink appears to be an exception since its paws have remained almost webless, a condition more typical of high-speed terrestrial runners. The limbs of mink are short and each of the five digits bears a curved, compressed, non-retractile claw. There is partial webbing between the toes, but little more than that found in the more terrestrial polecat. The surface area of the foot is relatively small suggesting that they are adapted for locomotion on land rather than underwater.

LOCOMOTION ON LAND

When walking, the head of the mink is held near the ground, the back is level and the tail is held taut and slightly arched towards its distal end. The sequence of limb movements is similar during fast and slow walking using an alternating gait whereby diagonal pairs of legs move simultaneously. The gait on land has been described as 'scampering', and is interspersed with a series of faster 'bounds'. The fast movement of the mink was designated full bound or half bound (Dagg 1973) depending on whether the limbs are simultaneously replaced on the ground or if one limb is replaced slightly before the other. Both versions include a phase when all limbs are off the ground and the forelimbs outstretched. During bounding the head is held high and the tail is taut and arched upwards. The sequence of limb movements during terrestrial locomotion is shown in Fig. 5.2. Speeds of 48 cm s^{-1} while walking and up to 262 cm s^{-1} during the bounding gait were recorded. This is broadly equivalent to five to six body lengths per second. During bounding, spinal flexion occurs which is absent during walking. With their elongate shape and

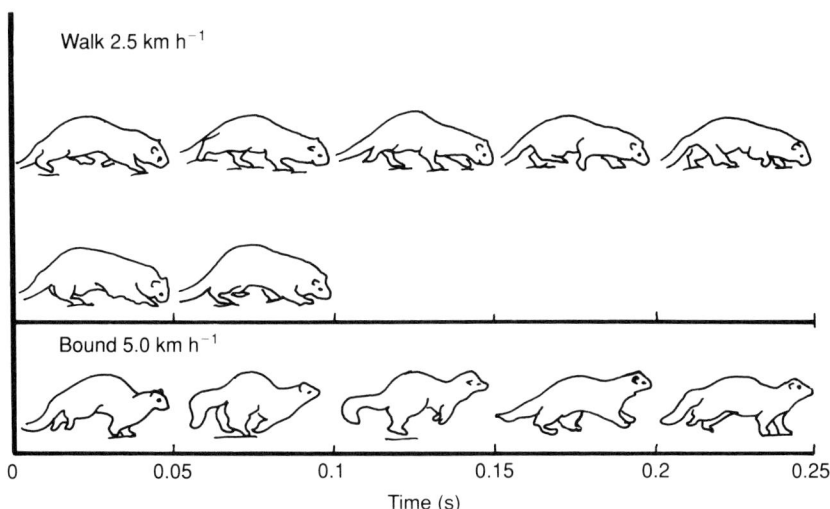

FIGURE 5.2 *Walk and half-bound gaits of mink. Source: Williams (1983b).*

short limbs, mink encounter problems of body support and stride length not experienced by longer-legged vertebrates.

Brown & Lasiewski (1972) consider the specialized shape of the mustelid body to have evolved as a hunting strategy which permits entrance into confined places in search of prey. The reduced limb length and increased body mobility necessary for navigating animal burrows does not appear seriously to limit terrestrial locomotion in the mink. Williams (1983a) has suggested that the increased flexibility of the spine has compensated for the disadvantages associated with short limbs by contributing to the efficiency of the bounding movement.

LOCOMOTION IN THE WATER

Two forms of aquatic locomotion are distinguished in the mink. When swimming fully submerged the mink uses a power and recovery method involving alternate strokes of either diagonally opposite legs or those on the same side of the body simultaneously (see Fig. 5.3). During the power stroke each limb scribes an angle of approximately 130° from the body and during recovery the flexed fore limb moves in a plane parallel to the body surface with the palmar surface of the paw facing upwards.

While all four paws are employed when swimming fully submerged, surface swimming is often achieved using the fore limbs only aided by an occasional thrust of a hind limb (Fig. 5.4). When swimming on the surface of the water the mink's head, the dorsal surface of the body and the posterior half of the tail are exposed.

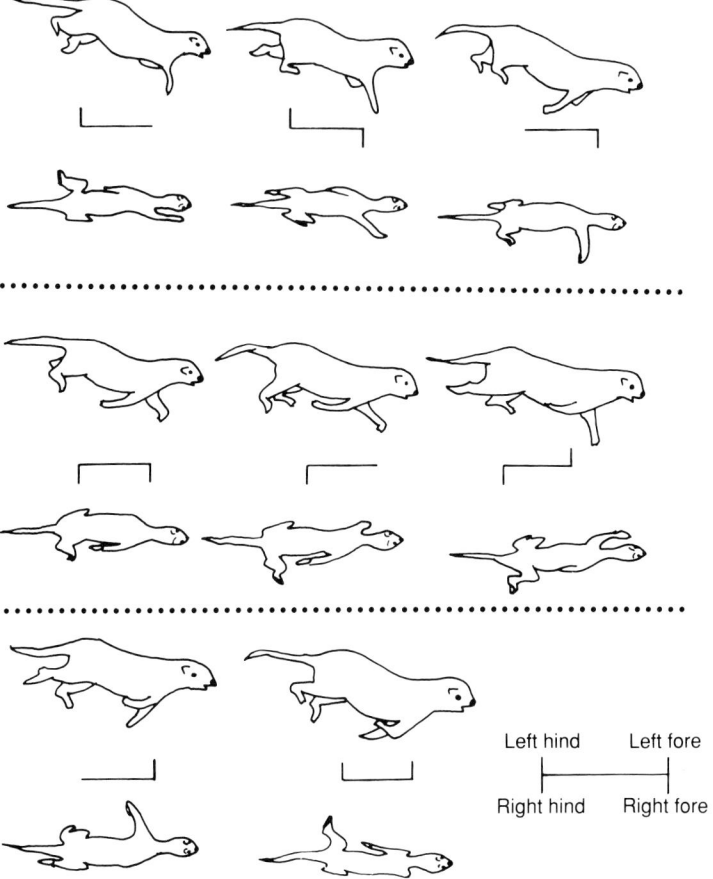

FIGURE 5.3 *Sequence of limb movements of the mink when swimming fully submerged. The bar diagrams show those limbs that are in the power stroke. Source: Dunstone (1981).*

Speeds ranging from $42\,\text{cm}\,\text{s}^{-1}$ while swimming on the surface to $59\,\text{cm}\,\text{s}^{-1}$ when pursuing prey underwater have been commonly recorded (Dunstone & Poole 1976). The top speed measured was $83\,\text{cm}\,\text{s}^{-1}$, approximately two body lengths per second. This is considerably slower than the top speeds attained by the otter of 10–$12\,\text{km}\,\text{h}^{-1}$ (or $300\,\text{cm}\,\text{s}^{-1}$), equivalent to four body lengths per second. Interestingly even the maximum speeds of the mink are slower than many of its fish prey. For example, Bainbridge (1958) has shown that a small dace (*Leuciscus leuciscus*) approximately 9 cm long was able to swim at $170\,\text{cm}\,\text{s}^{-1}$ and a 13.5 cm trout managed $225\,\text{cm}\,\text{s}^{-1}$.

Increase in speed is associated with an increase in stroke frequency from $2.7\,\text{strokes}\,\text{min}^{-1}$ when surface swimming to $3.7\,\text{strokes}\,\text{min}^{-1}$ when search-

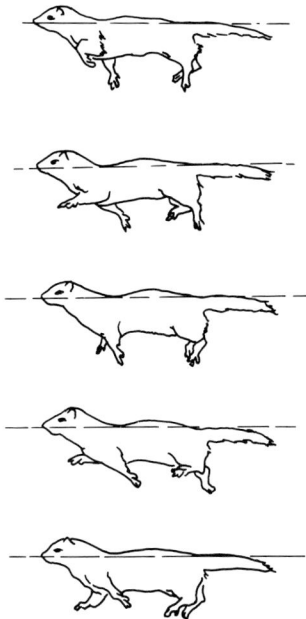

FIGURE 5.4 *Sequence of limb movements of the mink when swimming on the surface. Source: Williams (1983a).*

ing and 4.0 strokes min^{-1} when pursuing prey underwater. Mink diving into water from the tank edge use the hindlimbs to push off from the ground. The mink enters the water with its forelimbs held close to the body until the required depth is reached. When diving from the water surface, alternate strokes of the forelimbs are used to propel the animal downwards. There is no clearly discernable pattern for surfacing; generally the mink inclines its head upwards and floats to the surface. While swimming underwater the mink commonly makes use of solid structures as a base to push off from, giving additional propulsion. European mink are reputed to be even weaker swimmers than *M. vison* (Ternovsky 1977).

RESPIRATORY BIOLOGY

Mink are air-breathing mammals, and to search for prey underwater they must voluntarily suspend ventilation for an extended period. There are several adaptations which directly affect breath-holding capability during endurance diving in amphibious and aquatic mammals and birds. Firstly, the oxygen storage capacity of the body may be increased, both the gaseous reserves available in the lungs and those chemically-bound reserves of oxygen available in the blood haemoglobin and muscle myoglobin. Secondly, the tolerance of particular tissues to toxic chemicals may be modified. These chemicals include lactic acid, which builds up during exercise under conditions of low oxygen. Finally, the ability to restrict the peripheral blood supply to the skin by vasoconstriction allows sensitive tissues, such as the heart and brain, to maintain an adequate supply.

There is a considerable amount of information available on the physiological responses made by mink when diving. On land the mink breathes some 40–70 breaths min^{-1} although extreme variations in respiratory frequency are common. The heart of the mink beats at between 90 and 180 beats min^{-1} when the animal is at rest or undertaking moderate exercise. The beats are characterized by marked fluctuations in rate (arrhythmia). Fluctuations in the heart rate also occurred when the mink was frightened or when it investigated a novel object.

Early studies of the physiology of diving mammals involved forced submersion as the main technique of study. Under these circumstances it was generally observed, probably not unexpectedly, that the heart rate dropped dramatically when the animal was kept underwater for a protracted time. This is referred to as a diving bradycardia. Such a phenomenon has also been observed in mink when they have been forced to undertake dives (Gilbert & Gofton 1982a, West & Van Vliet 1986). Diving bradycardia has been interpreted as an oxygen conserving response. However, reduction in heart rate alone would not be sufficient to allow an animal to remain submerged for extended periods, and it may be just one of several physiological responses that occur when diving underwater. It is, of course, more appropriate and humane to monitor the behaviour of freely diving, unrestrained animals. Studies conducted by Stephenson et al. (1988) successfully monitored the heart rate and oxygen consumption of mink using small implanted radiotransmitters during normal foraging dives in a large respirometer in the laboratory (see Fig. 5.5). When swimming in a familiar tank at the University of Durham where the animals had been trained they did not show any reduction of heart rate during normal shallow (0.3 m) or deep (1.9 m) dives. Interestingly, when the animals encountered a novel situation, or were diving in an unfamiliar tank at the University of Birmingham, they did show a reduction in heart rate suggesting that this may be associated with a fear response.

Figure 5.6 shows a recording made of a mink's heart rate (electrocardiogram) while it was undertaking a series of dives. The heart rates of these freely diving mink are highly variable (Fig. 5.7). The highest rates observed during voluntary dives in a shallow pool at Durham were between 210 and 306 beats min^{-1} compared to between 200 and 208 beats min^{-1} when the animals were at rest in their home cages. At the other extreme, the lowest heart rates while diving were recorded as the mink entered and explored novel objects such as submerged pipes, when the heart rate then fell to 56–83 beats min^{-1}. This variability of heart rate response suggests that psychological factors can have a potent effect on the physiological responses of the mink.

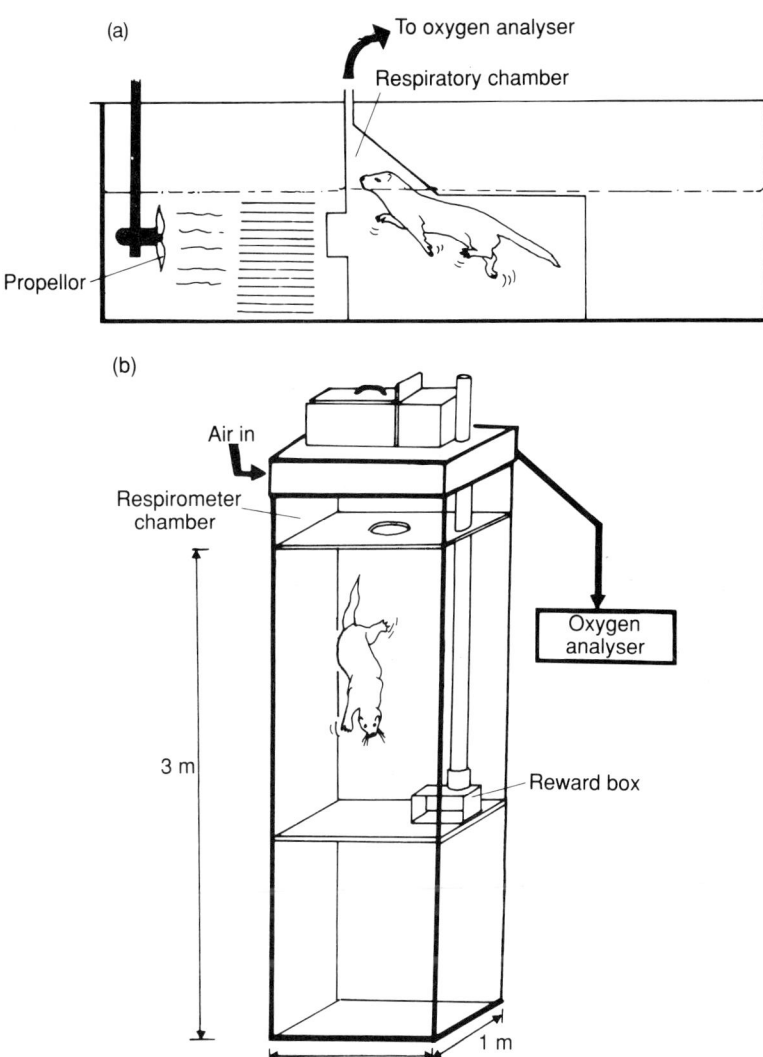

FIGURE 5.5 *Apparatus for measuring energy expenditure in freely-diving mink: (a) flume, (b) deep-diving tank.*

Another observation which indicates that psychogenic influences may affect heart rate is that a pronounced anticipatory increase in heart rate often occurred before the mink broke surface at the end of deep dives.

Other experiments on the heart rate of freely diving mink carried out by Gilbert & Gofton (1982a) suggested that bradycardia may also develop if the animals are frightened into diving. Monitoring the heart rate of drowning

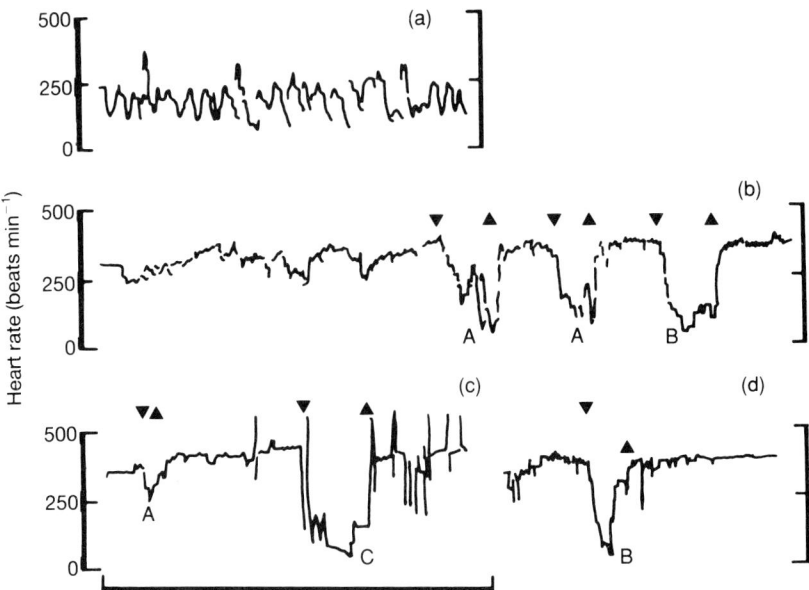

FIGURE 5.6 *Electrocardiogram (ECG) of freely-diving mink: (a) at rest, (b) in a novel tank, (c) in a familiar shallow tank, (d) in a familiar deep tank. Arrows indicate direction of movement; A, shallow dives; B, deep dives. The scale bar below the traces represents one minute. Source: Stephenson et al. (1987).*

mink caught in leg-hold traps, Gilbert & Gofton (1982b) noted that submergence was followed by an immediate reduction in the heart rate from 227 beats min^{-1} to 119 beats min^{-1}. Death by drowning resulted after 4.5 min. Subsequent post-mortem showed the mink to have died from wet-drowning, probably brought about by a sensitivity to the build-up of carbon dioxide in the blood leading to the breathing reflex being triggered and water being inhaled.

For many aquatic species an adequate supply of oxygen should be available in the storage capacity of the lungs, blood and muscles for the entire dive. This is particularly likely to be the case for shallow diving animals that remain submerged for only a short time. Laboratory estimation of oxygen storage capacities and oxygen consumption rates by respirometry have shown that most animals dive for a shorter period than that which would exhaust their stored oxygen reserves and cause them to switch to anaerobic metabolism. This suggests that anaerobic metabolism may be reserved for emergency situations.

Although lung volume increases with body size across many mammal species, there is no evidence to show that the lung volume of amphibious mammals is proportionately larger than that of their terrestrial relatives. It is

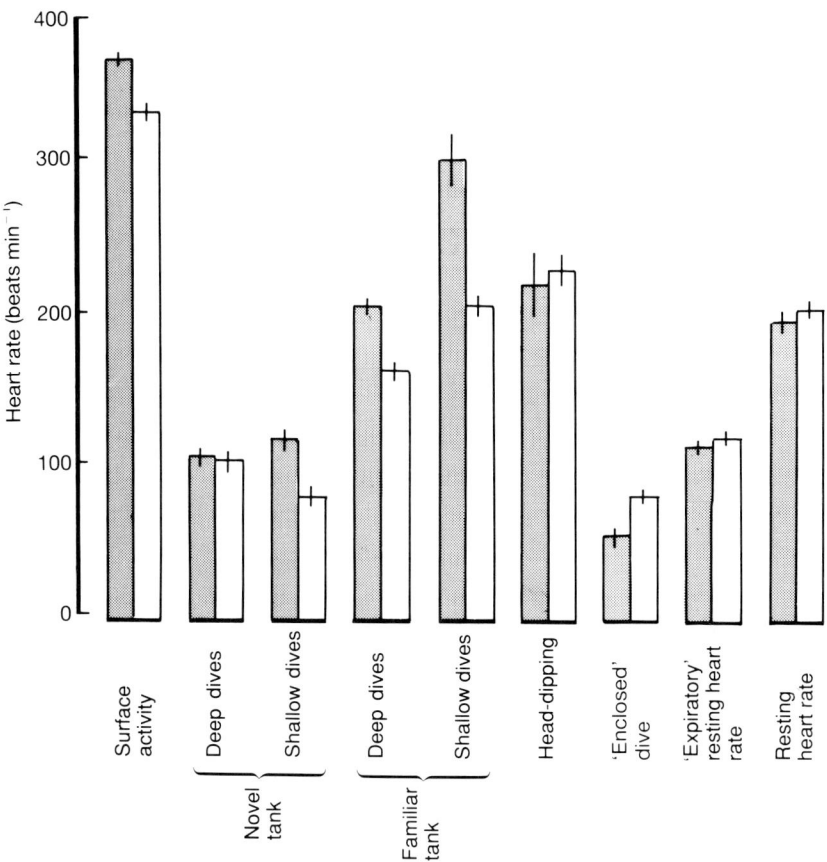

FIGURE 5.7 *Heart rate of mink. Source: Stephenson et al. (1987).*

not known whether mink have evolved specializations to increase their oxygen carrying capacity, although some preliminary tests I have conducted on mink blood do suggest that the haemoglobin concentration may be higher in the mink than in its terrestrial cousin, the ferret.

In conclusion, it would appear that the normal foraging dives of the mink are aerobic, that is, carried out using the body's normal oxygen reserves. The possibility that the animals are consciously able to initiate the development of a bradycardia cannot be excluded if the situation (e.g. sustained pursuit of a fish or escape from a predator) requires it. The maximum dive time I have recorded for mink has been 30 s, although dives are commonly much less than this and in fact most are less than 10 s. But how do mink decide when to terminate a dive and how can they extend dive times physiologically when circumstances require it? It is likely that normal shallow dives are within the aerobic diving limits of the mink and that other constraints, such as those imposed by foraging strategy, act to keep dives even shorter.

ENERGETICS

THE COSTS OF BASAL METABOLISM

The 'tick-over' energy requirement of an animal at rest, the so-called basal metabolic rate (BMR), increases with body weight in a fairly predictable manner within and across a wide range of species. The elongated shape of mustelids, a design for the capture of prey in confined spaces, must constitute a very successful evolutionary strategy, as it has been retained by all members of the group, despite their diverse range of habits, over considerable evolutionary time. But it has had profound consequences for much of the mink's behaviour and lifestyle. For mustelids weighing 1 kg or more, the BMR was 20% higher than that predicted (Iversen 1972). He suggested that the elongated body shape contributes to this higher resting metabolic rate since the animals cannot curl into a heat conserving spherical shape when resting and because of the great heat loss resulting from the high skin surface area to body volume ratio.

THE COSTS OF LOCOMOTION

Heart rate and oxygen uptake can be remotely monitored during diving activities using the techniques of radio-telemetry and respirometry (Stephenson *et al.* 1988). In this study a female mink which had been surgically implanted with small heart rate transmitters was encouraged to dive for pieces of fish flesh presented at a given depth on the floor of a tank. Heart rate was also measured during head-dipping when only the head and thorax are voluntarily submerged as the mink searches for its prey, and during shallow and deep dives. The costs of locomotion can be estimated by determining rates of oxygen uptake (V_{O_2}) and carbon dioxide production (V_{CO_2}). This is relatively easy to accomplish for terrestrial animals by training them to run on a treadmill moving at a particular speed (see Fig. 5.8). The

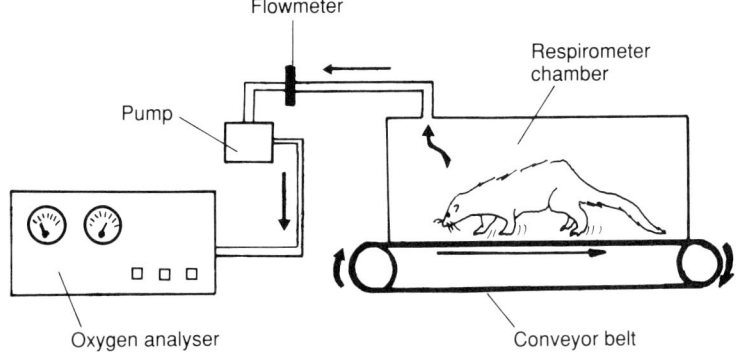

FIGURE 5.8 *Measuring oxygen consumption from a mink on a treadmill.*

equivalent technique for aquatic locomotion is to train the animal to swim in a water flume, an artificially created water current running at constant speed (see Fig. 5.5a). Stephenson *et al.* (1988) obtained values for V_{O_2} of approximately 5 ml O_2 g h^{-1} for a 650 g mink and 8 ml O_2 g h^{-1} for a 1000 g mink when swimming in water. Williams (1983a,b) measured the oxygen uptake of a female mink surface swimming at 0.7 m s^{-1} or running on a treadmill as being broadly equivalent at 6.5 ml O_2 g h^{-1}. The energy costs were smaller for males. Surface swimming is a very energy demanding process and it is estimated that the mink achieves a five- to ten-fold reduction in costs when it swims underwater rather than on the surface. This was due to reduction in the drag of the poorly streamlined mink body as it ploughs through the water. The mink is a very inefficient swimmer as the small surface area of the paws does not provide efficient thrust generation, and the alternating use of the paired limbs is not as effective compared with the synchronous thrusts produced by the otter. The main diving strategy (see Chapter 6) of the mink appears to involve locating prey from an aerial vantage out of the water, then diving onto the prey with the extra velocity provided by pushing off with the hind legs (dive chases). Hence, the mink reach their prey more quickly and with less energy expenditure allowing more of their limited dive time for pursuit underwater.

The increase in hydrostatic pressure associated with diving to depth may also be important since this will compress the air in the lungs, leading to increased uptake of lung oxygen stores by the blood. The reduced lung volume would also reduce buoyancy, further lowering the cost of locomotion, since the mink will have to expend less energy staying submerged at a particular depth. Dive-chases may assist the mink in achieving a depth while maintaining a generous volume of oxygen in the lungs.

SENSORY BIOLOGY

Terrestrial mammals are probably less specialized in the use of their senses than other vertebrate groups. The stimuli most useful to a predator depends on the characteristics of the prey species being hunted, the habitat hunted over and the time of day at which hunting occurs. Of the limited number of mammalian species that have been investigated experimentally, the trend appears to be for vision to be used by diurnal hunters, while audition and olfaction are more important to nocturnal predators.

SENSES USED DURING PREDATION

There have been several descriptions of the predatory behaviour of terrestrial mustelids including the weasel (*M. nivalis*), the stoat (*M. erminea*) and the polecat (e.g. Eibl-Eibesfeldt 1956, Wüstehube 1960, Apfelbach 1973). All place importance on the senses of vision and olfaction for the detection and pursuit of prey. Usually it has been shown that movement of the prey is sufficient to release chasing by the predator, although smell is probably the more important stimulus in eliciting such behaviour. The predatory sequence of the mink hunting on land is likely to be similar to the polecat, but has yet to be studied in detail.

Sensory input is required at each stage of the predatory sequence, but particularly during the initial detection and recognition of the prey. Müller (1930) considers that for mustelids the senses can be ranked in the following order of importance: olfaction, audition, and finally vision. However, more recent studies have suggested that visual cues are important in causing the predator to orientate towards its prey. In a study of the sequencing of predatory behaviour in the weasel, stoat and polecat, Wüstehube (1960) found all three predators to be more reliant on visual cues from the prey than on scent stimuli. While mink probably resemble their terrestrial relatives when hunting on land, the considerable contribution of aquatic prey to their diet necessitates different sensory adaptations.

Hunting underwater

> Greeting water as a new friend or an old enemy?
>
> Walls (1965)

The basis of Walls' remark is that our vertebrate ancestors originally evolved in water, and presumably had sensory adaptations enabling them to use that environment. Present day mammals which use the seas, rivers and lakes for food are secondarily adapted to this habitat.

When one considers the special case of an amphibious mammal hunting underwater, vision becomes to some extent obligatory, even though low contrast and absorption of light underwater will impair vision at anything but close range. Whereas olfaction and audition may be of importance to mink hunting on land, these senses are of limited use underwater. Tactile information can provide a useful alternative in very close encounters, or may act as a substitute when murky water precludes the use of vision.

An amphibious predator hunting by vision would be expected to possess equivalent visual capability in air and underwater. A useful measure is the ability of the animal's eye to resolve fine detail—its visual acuity. This is measured as the minimum angle subtended at the eye by the smallest detail that the eye can resolve. The units of measurement are minutes of arc (one minute is one-sixtieth of a degree). Hence good visual acuity is indicated by small visual angle. Man is considered to have good visual acuity with a visual angle of 0.5 min of arc.

It is the general case that predators require acute vision to detect and then correctly identify prey. In contrast, prey species lack good acuity in favour of wide angle vision and sensitivity to low light levels (Walls 1965). This fundamental difference between predators and their prey is demonstrated in Table 5.1. In wholly nocturnal predators there is a trade-off between visual acuity and sensitivity to dim illumination. This has resulted in slightly poorer acuity than might otherwise be expected; for example, the cat has a visual

TABLE 5.1 *The visual capability of nocturnal and diurnal animals.*

Species	Acuity (min of arc)	Reference
Nocturnal		
Ferret/polecat	16.2	Neumann & Schmidt (1959)
Cat	5.5	Smith (1936)
Rat (pigmented)	26.0	Lashley (1930)
Rat (albino)	52.0	
Diurnal		
Human adult	0.44	Walls (1965)
Rhesus monkey	4.0	Walls (1965)

acuity of only 5.5 min of arc because it hunts mainly at night. Mustelids in general have been found to possess considerably poorer acuity than the cat, and this is probably due to the greater importance of olfactory stimuli to these species.

Estimating visual capability

Psychophysical experiments have been carried out to determine the visual acuity of mink in air and underwater over a range of viewing distances and ambient lighting levels (Sinclair *et al.* 1974, Dunstone & Sinclair 1978). The results of these studies show that visual acuity for static objects declines markedly underwater, from 15.1 min of arc in air to 31.4 min of arc when underwater at moderate illumination levels. When ambient illumination was lowered, the visual acuity was further reduced to 51.7 min of arc in air and 95 min of arc underwater. This suggests that if mink confine their underwater hunting to brighter light levels than they should be no worse off than when hunting on land under dim light conditions. As the distance from which the mink viewed the stimuli was increased, visual acuity deteriorated in air. This happens at an even faster rate underwater. Overall the results suggested an optimum viewing distance underwater of between 15 and 60 cm. This correlates with the average detection distance for mink detecting fish prey underwater of about 35 cm (Poole & Dunstone 1976).

Since it is generally accepted that mustelids hunting on land rely heavily on the detection of prey movement, Clements & Dunstone (1979) undertook a series of experiments to investigate the ability of mink to detect high-speed directional movement. The stimulus used was a small spot of light projected onto a cathode ray tube. This could be programmed to move horizontally over a wide range of speeds. The mink was required to respond to the direction in which the spot was travelling. The most interesting thing to note was that motion detection appears not to be deleteriously affected by submersion underwater since broadly equivalent results were obtained in both media (Clements & Dunstone 1979).

ANATOMY AND PHYSIOLOGY OF THE EYE

The eyes are quite frontally positioned on the head and this provides for a binocular visual field of 80°. By comparison, the even more forward-facing eyes of the cat provide for a considerably greater binocular field, perhaps as much as 130° (Walls 1965).

A sectioned eye of the mink is shown in Fig. 5.9. The light receptor cells in the retina of the vertebrate eye are classified into two groups according to their function and histological appearance. The cones are small slim units responsible for colour vision, but work only when bright light is available. The rods are characteristically interconnected into large functional units. These rods are sensitive to low light levels, but do not possess any colour sensitivity. For the visual system of any vertebrate, sensitivity and acuity are determined by the presence or absence and relative proportions of rods and cones in the retina. Interconnection between rod cells allows for summation of sensory input and therefore high sensitivity to low light levels. However, this process causes rod vision to be diffuse. The independent responsiveness

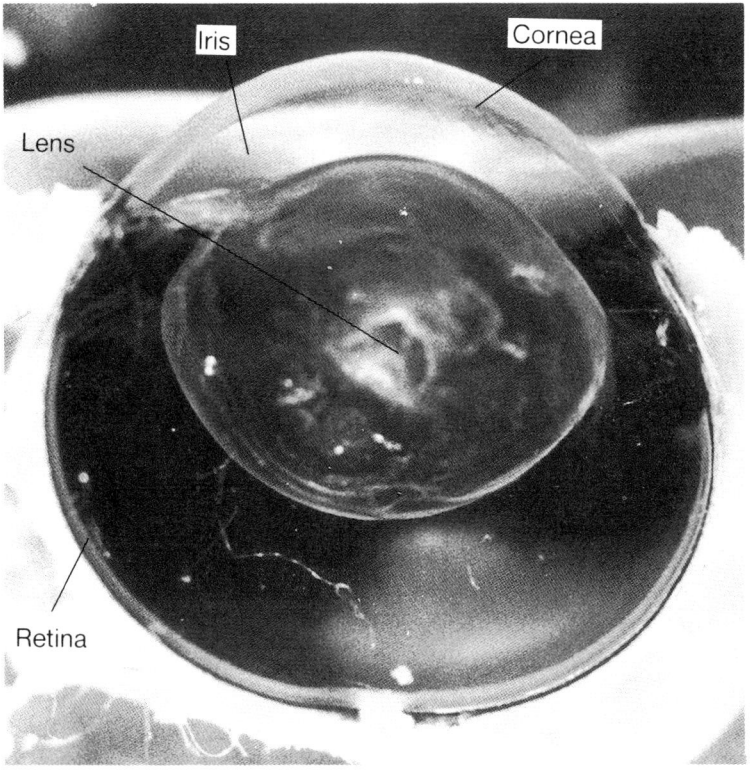

FIGURE 5.9 *Section of the eye of a mink. Photo: Nigel Dunstone.*

of single cones allows for sharp vision, but at the expense of sensitivity since they are not usually interconnected.

Vertebrate eyes may vary from those containing entirely rods in some nocturnal species or fossorial species such as the mole, to entirely cones in daytime active species such as the squirrels. There is a general consensus that mustelids have a duplex retina, that is, containing both rods and cones, but with rods predominating. The ratio of rods to cones in the mink has been measured as 20:1 at the ora serrata, the most sensitive part of the retina (Herter & Klaunig 1956, Dubin & Turner 1977) (see Fig. 5.10). This high

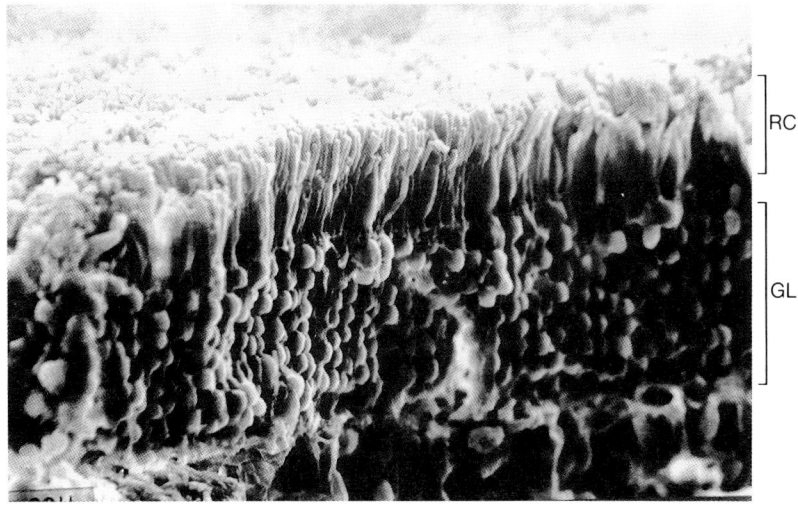

FIGURE 5.10 *Scanning electron micrograph of the retina of the mink showing rods and cones. RC, receptor cells; GL, ganglion layer. Photo: Nigel Dunstone.*

density of rods indicates that it is essentially a nocturnal retina. A well-developed reflective layer called the tapetum is also present giving the eye a blue-green eyeshine when illuminated at night. This device serves to improve sensitivity to low light levels by acting as a mirror behind the retina: the tapetum reflects the light back out of the eye, so stimulating the sensory cells a second time. Under dim light conditions, as the pupil becomes larger to allow in more light, optical defects of the cornea and lens (particularly spherical aberration), become prominent adding to the limitations on vision caused by lack of photoreceptor sensitivity.

Mustelids vary in their degree of diurnality, and consequently might be expected to vary in the extent to which they possess colour vision. Walls (1965) mentions that mustelids, in particular the mink and the polecat, are sensitive to red; however, Müller (1930) failed to demonstrate this using behavioural tests in which animals were required to discriminate between coloured stimuli. In 1959, Gewalt showed that several species of mustelid including the mink have the ability to distinguish colours from brightness-matched shades of grey. In these more rigorous tests, mink were able to discriminate red and green, and also showed some sensitivity to yellow and blue.

The intensive breeding and selection of mink for fur colour has sometimes led to colour varieties being developed with a loss of body pigmentation, that is, albinism. In some cases this has also involved a loss of retinal pigment and the development of abnormal visual pathways and presumably inferior vision.

Adaptations for vision underwater

Vision underwater is complicated by the similar refractive indices of the cornea and water. As a result, the cornea can no longer contribute to the focusing power of the eye. Since in terrestrial animals the cornea is the principal refracting surface of the eye, under water such an eye would focus the image behind the retina resulting in blurred vision (see Fig. 5.11). Submergence thus causes longsightedness (hypermetropia). To maintain its acuity in water, the eye of an amphibious mammal must have greater focusing or dioptric power. In optical terms this means a higher lens curvature.

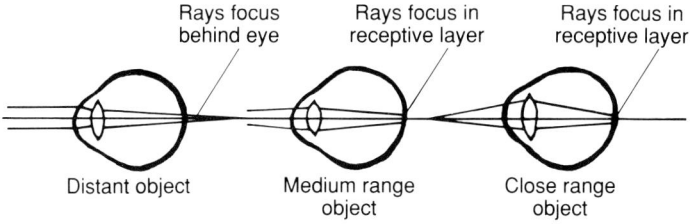

FIGURE 5.11 *The mechanism of hypermetropia: left, viewing a distant object; middle, viewing a medium range object; right, viewing a near object.*

Aquatic mammals have evolved a variety of mechanisms to overcome this problem. Table 5.2 gives the comparative visual acuity data for a number of species of amphibious and aquatic animals that have been tested. Whales, dolphins and porpoises have adopted the 'fish-eye lens' solution by evolving a spherical lens of high curvature and hence high dioptric power. However, this causes problems when these animals, albeit rarely, have a requirement to see in air. Seals and sea lions also use this method, but manage to achieve reasonably acute vision in air by using a narrow pupil which functions like a pin-hole camera. As a result their aerial vision deteriorates markedly with decreasing light levels. Another approach has been adopted by the cormorant and otter and involves using a well-developed sphincter iridis muscle to compress the outer edge of the lens to produce an area of high curvature and hence powerful focusing ability (see Fig. 5.12). Using this method the otter is

TABLE 5.2 *The visual acuities of various species of aquatic and amphibious mammals.*

Animal	Visual acuity (min of arc)		Reference
	Air	Water	
Steller sea lion *Eumetopias* sp.	7.0	—	Schusterman & Balliet (1970a)
Harbour seal *Phoca vitulina*	8.3	—	Schusterman & Balliet (1970a)
California sea lion *Zalophus californianus*	—	5.5	Schusterman & Balliet (1971)
Dolphin *Lagenorhyncus obliquidens*	—	6.0	Spong & White (1971)
Killer whale *Orcinus orca*	—	5.5	White *et al.* (1971)
'Clawless' otter *Amblonyx cineria cineria*	13.6	—	Balliet & Schusterman (1971)
Mink *Mustela vison*	15.1	31.4	Sinclair *et al.* (1974)

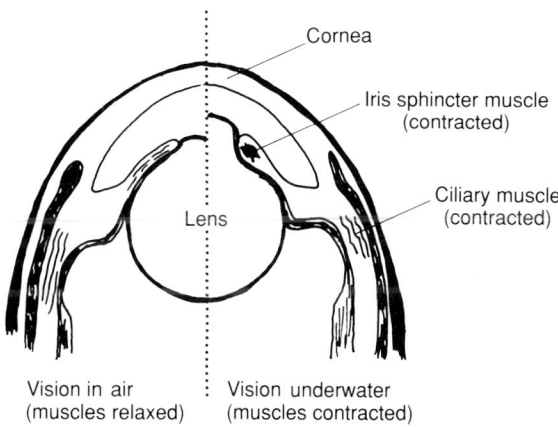

FIGURE 5.12 *The mechanism of accommodation in the otter.*

believed to achieve equivalent levels of visual acuity in air and water, at least in bright light. Histological studies on the mink eye suggest that it may also employ this method (see Fig. 5.13). Certainly the development of the iris sphincter muscle of the mink is greater than in the ferret but possibly less than in the otter.

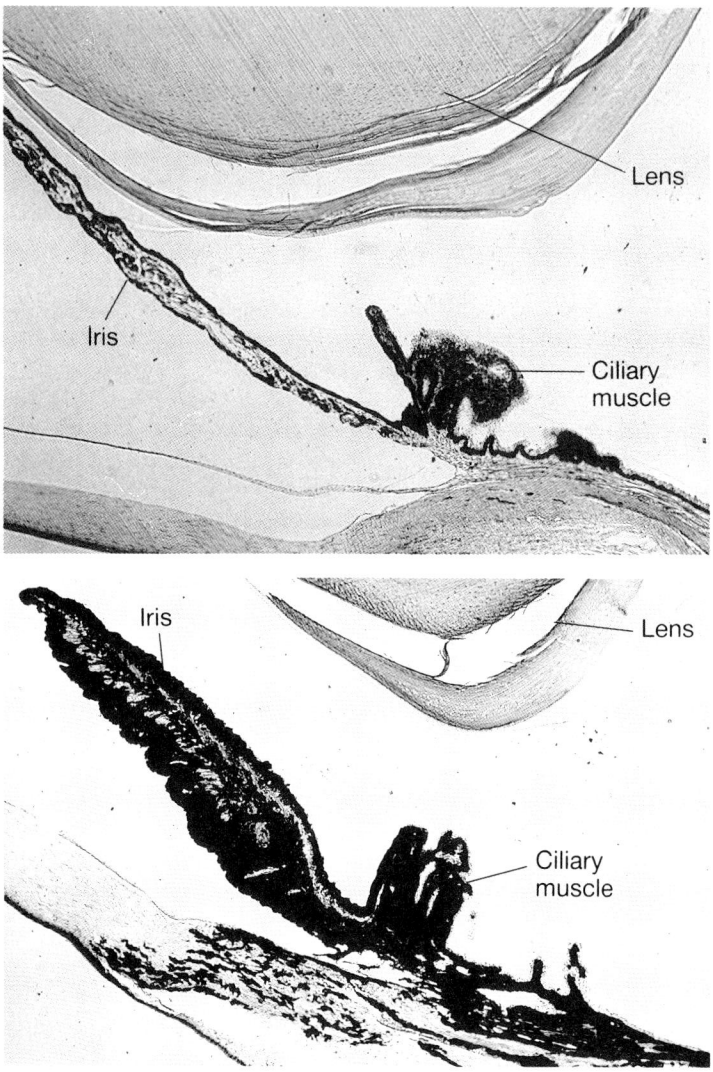

FIGURE 5.13 *Development of the sphincter iridis muscle in the ferret (top) and the mink (bottom). Photo: Nigel Dunstone.*

HEARING AND VOCALIZATIONS

The basic sound repertoire of the mink has been analysed by Gilbert (1969) using a sound spectrograph. This machine translates the animal's sounds into a graph of pitch (frequency measured in kilohertz, kHz) against duration. It is usual to stretch out the time axis so that analysis of very brief vocalizations is possible (see Fig. 5.14). Defensive and warning sounds produced by mink

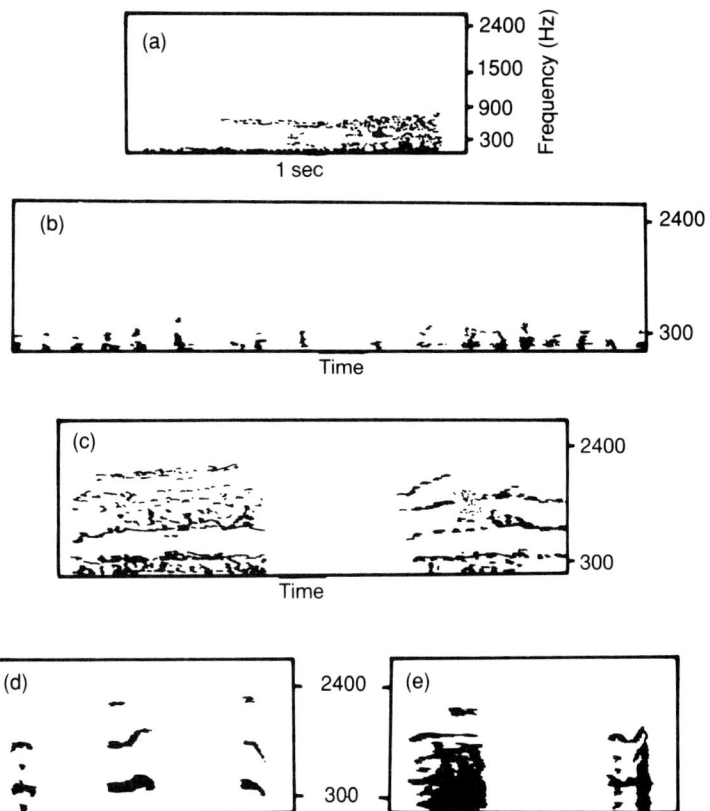

FIGURE 5.14 *Vocalizations of the ranch mink: (a) male hiss, (b) male chuckling, (c) female scream, (d) female aggressive squeaks, (e) male scream and squeak. Source: redrawn from Gilbert (1969).*

include a hiss (duration 0.8 s, frequency 0.6 kHz), which is a low-intensity aggressive response to threat. It is a sound predominantly made by the male. The scream is a complex sound of variable duration (mean 1 s) used as a defensive, and very intimidating, threat vocalization. Its maximum intensity is about 20 decibels (dB) and the fundamental frequency rises very sharply from 0.15 to 0.325 kHz with overtones to 2 kHz. Squeaks are usually associated with pain or fear; these are of short duration (0.05–0.3 s) and often emitted in a stacatto burst. Often there is a sharp rise and fall in pitch from the strongest frequency present of 0.65 kHz.

It is reasonable to assume that the mink is capable of hearing the vocalizations it makes, but how much higher does its frequency reception extend? Hefner & Hefner (1985) have shown that the peak sensitivity of hearing for many carnivores, including the closely related weasel is in the range 1–16 kHz. Recently Powell & Zielinski (1989) have demonstrated that mink are also able to hear tape-recordings of the ultrasonic calls emitted by their rodent prey at 40 kHz.

OLFACTION

Most mammals have a highly developed olfactory sense. The use of chemical signals has a number of advantages over other means of communication. Compared to visual and auditory signals, scent is long-lived and thus has the potential to provide both spatial and temporal information. By this we imply that the chemical signal indicates by its very presence that the producer has been at this location. Additionally, presumably by virtue of its decay characteristics, information will be available concerning how long ago the scent was deposited. Like auditory signals, scent can be used at night.

The detection of the scent of prey may also be crucially important during hunting. In addition, smell is extensively employed in social communication, using chemical signals originating in urine, faeces or cutaneous scent glands. Even solitary animals, like the mink, have a need to communicate with neighbouring territory holders.

In mink, the anal gland is a major source of specialized odour compounds (Brinck *et al.* 1978, 1983). This term is used for all glandular tissue around the anal region, including the anal sacs. The anal glands of mustelids comprise two anal pouches, each connected with a complex of cutaneous glands. Within the anal sacs are located blocks of apocrine and sebaceous tissue. The secretion from the anal sacs is discharged through ducts just internal to the anus.

The chemical compounds contributing to scent fall into two categories. The first category comprises highly volatile, low molecular weight compounds, whose rapid dispersal is used in alarm signalling. The release of such strong odours when the animal is frightened may act as a warning to conspecifics and also as an anti-predator mechanism. Secondly, a group of higher molecular weight compounds provides a more persistent scent. They impart individual odour differences to the animals probably making it possible for discrimination between individuals of their own species. Brinck *et al.* (1978) suggest that, in a stable territorial system, this will allow the acknowledgement of the presence of neighbours without confrontation while still allowing territorial intruders to be detected. No sex differences were detected in the analyses of anal sac secretion, but these may be provided by chemicals in the urine which often accompany scent marking.

Marking is performed in several different ways. Two of the most common are anal gland dragging by rubbing the anal region against objects on the ground, and deposition of faeces. During anal dragging the animal adopts a sitting posture bringing its anus into contact with the ground. It then proceeds to shuffle forward, pulling itself along with its forelegs and dragging its anus against the ground. Presumably this action brings about deposition of the contents of the anal sacs. During defaecation the mink sniffs at an appropriate spot, turns to face away and backs up over the chosen area. It then arches it's back, lifts it's tail and deposits a scat, often while simultaneously urinating. The sequence is frequently followed by anal dragging. This behaviour is particularly prevalent when the mink is introduced into a new area or when it emerges from the water after swimming.

Scent trails may be also deposited during movement, even in the absence of any obvious marking behaviour. We know these are there because other mink follow them like a hound coursing a fox. Urine trails of up to 0.5 m long are occasionally deposited by the moving mink. Again the depositing of such a trail is not easy to observe, but sometimes mink are seen to squat and waddle with the tail raised.

Gerell (1968) observes that faeces are often deposited on prominent, elevated features in the territory, for example rocks and stumps, possibly to enhance the active range of the scent. Such sites are scattered throughout the mink's territory, and there is no reliable evidence that boundaries receive special attention. The faeces have a strong odour mainly originating from the secretions of the proctodeal glands. In mink, defaecation and urination often accompany scent-marking. Scent marks of other mink often stimulate over-marking by the territory owner.

All mustelids are able to empty their anal sacs by contracting the muscles around the pouches. Occasionally this is seen as a spray of clear liquid. More often all that is detected is the smell. A mink readily displays this characteristic, particularly when threatened: the mink bristles its fur, arches its back and emits the foulest of odours.

A different set of glands, presumably located in the skin of the chest, is employed when the mink rubs its ventral surface on the ground. This is commonly observed when the mink emerges from the water after swimming, and when it encounters other mink. The mink will lie on its belly with the entire ventral surface in contact with the substrate. The chest and throat are rubbed against the ground in a characteristic back and forth motion.

Experimental evidence has been provided by Robinson (1987) that mink are well able to discriminate their own faeces from those of other mink. In a series of painstaking tests he further demonstrated that mink can distinguish between scats of known and unknown mink of the same sex, and between unknown males and females. Further experiments demonstrated that urine provided information regarding the reproductive status of the female.

VIBRISSAE

Vibrissae are specialized tactile hairs which have a rich nerve supply in comparison to other hairs on the body. They are an important source of information for many mammals, especially those with crepuscular or nocturnal habits. In the mink they are not confined to the snout, but are also found above the eyes and on the ventral surface of the neck and limbs. Fig. 5.15 shows the vibrissae in use during fish capture. Like other hairs they consist of a capsule, the follicle, embedded in the skin epithelium, but unlike a

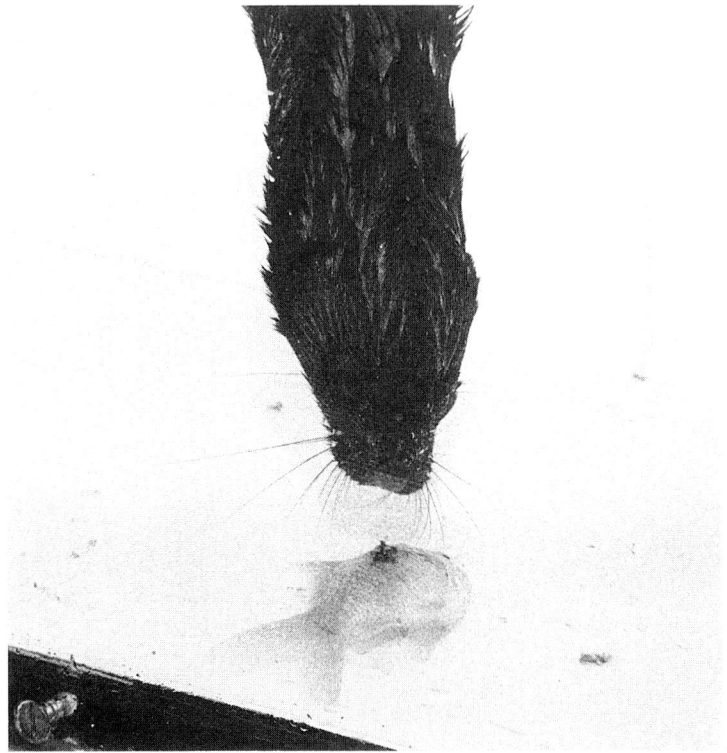

FIGURE 5.15 *The facial vibrissae in use during fish capture. Photo: Nigel Dunstone.*

normal hair follicle they are innervated by four types of receptor cell. This provides for extreme sensitivity, to detect contact, but also the extent and direction of bending. It is likely that they provide information to the mink on its velocity as it moves through its habitat. Fine tuning of movement at close quarters is also afforded, and may be particularly important in capture and manipulation of prey. Green (1977) has shown that otters whose vibrissae have been trimmed show a reduced efficiency at catching fish prey. The otter uses its vibrissae to detect underwater prey movement, probably caused by turbulence. Like the otter, the mink's muzzle is surrounded by stiff whiskers. The mink's reliance upon these for the detection of prey in murky water and at night is presently being investigated in a series of laboratory experiments.

In a similar series of experiments, with mink searching underwater in the dark for submerged plastic objects, I have shown that mink take longer to locate objects when their vibrissae have been shaved off. The ability of mink to distinguish between objects of different shapes was also impaired when vibrissae were absent.

ECOLOGICAL IMPLICATIONS OF BODY SIZE

The size of an animal has numerous implications for its behaviour and biology. This arises because the cost of upkeep varies according to size, and because growth rate cannot be scaled up in proportion to body size. Even when totally inactive a mammal must expend energy to maintain its constant body temperature. Depending on the level of activity exhibited, up to 80–90% of energy consumption may be used in this way. Body heat is lost through the skin. As a small mammal grows larger its volume increases faster than does its surface area. Consequently a larger mammal has less skin per unit volume and should lose heat more slowly. Because of this, larger mammals have a smaller energy consumption relative to their body size than do small mammals. As a result, smaller animals generally have high metabolic rates to cope with the demand of replacing heat loss. Due to their lower metabolic rate, large mammals cannot grow as fast, and hence their embryos require a longer period of development, both in the womb and after birth. The cost of this extended period of parental care tends to be minimized in larger species by the production of smaller litters. In general, this means that the potential for population growth in large mammals is considerably less than that in smaller species. Conversely, mammal species with high metabolic rates have greater capacities for rapid production of young.

Those species with a high reproductive potential have a great advantage in unstable environments as they can breed prolifically and at short notice to take advantage of any sudden abundances in their food. The key feature of these unstable environments is that sometimes food supply may exceed demand. Animals can respond by increasing their population relatively free from competition. It is implicit that at other times a food shortage will occur, and competition will be so high that many perish. Thus, those that have produced the greatest number of offspring have the greatest chance of having their genes represented among the survivors. Species adapted to these conditions are said to be *r-selected*.

In a stable environment the population density is finely adjusted to what the environment can sustain. Competition for what is available will be very intense. The smaller litters and protracted development times characteristic of the larger mammals are more suited to this case. It means that these species can invest more in the quality of their offspring. Such animals are said to be *K-selected*.

Competition leads to adaptive radiation and biodiversity. When all the usual resources are fully exploited, the animal who is able to satisfy its needs with an alternative resource less popular to its fellows, gains an advantage by reducing competition. Gradually this leads, through evolution of adaptations, to increasingly marginal and thinly scattered resources being used, and more severe environments and climates being tolerated. Thus, large body size can confer an advantage, with the ability to subsist on less frequent meals, to travel farther and to survive colder temperatures.

EVOLUTION OF SEXUAL DIMORPHISM

A number of species of vertebrates show sex-related differences (dimorphism) in body size. In a discussion of body size in a variety of Soviet mustelid species, Shubin & Shubin (1975) stated:

> ... ecological differences between the females and males are so great in these animals, particularly as regards their methods of obtaining food, their behaviour, and their activity, that they are as different from each other as absolutely different species.

Sexual dimorphism of body size is a characteristic feature of mustelid carnivores, particularly those in the subfamily Mustelinae. Males are always the larger sex and may, in some species, be more than twice the size of females. Sexual dimorphism ratios are found to vary between species of mustelines, and within the same species as a function of geographical range. Generally, the farther north you go, the greater the difference between the sexes.

The sexual dimorphism ratio is nearly always described on the basis of body weight. In mink, males are commonly 1.75 times as heavy as females. However, more than an increase in mass is implied. Indeed, sometimes it may be the concomitant increase in size *per se* that is the important difference between the sexes. Nevertheless, weight is more easily measured in the field. Dimorphism is evident in shape as well as size. For instance, the shape of the skull differs in several respects between male and female mink.

Several theories have been advanced to explain the sexual dimorphism of body size. These theories fall into two main groups. Those in the first group are based on the idea that sexual dimorphism has arisen to reduce competition between males and females by allowing them to exploit different types of prey (Brown & Lasiewski 1972). The second group of explanations concern the differing reproductive roles of the male and female; in particular, the selective pressures promoting large size in males and/or small size in females (Erlinge 1979, Moors 1980).

Reduced competition theories

To take the argument for the first theory, sexual dimorphism might be selected for as a means of reducing competition for available food resources. Since prey density is inversely related to prey size, predators specializing in small prey types benefit through having a more abundant food supply. Conversely, concentrating on the largest prey the mink is able to overpower gives a greater reward for hunting effort. In this case we might expect that a relationship would exist between the body sizes of predator and prey. Although a positive correlation has been demonstrated between the two, prey size is not a good predictor of predator size.

Predator size is governed not only by the size relationship between predator and prey, but also the presence of sympatric predator species which may share the same habitats and prey resource. Their co-existence may be made possible because of a difference in the sizes of the species. Such groups of closely related predators sharing the same resources have been referred to as 'hunting sets' (Rosenweig 1966, 1968). Similarly, competition between individuals of a species might be reduced when the sexes differ in size. In those parts of a species' range in which potential competitors were absent, McNab (1963) found body size to increase so that the adjacent niche could be filled; that is, larger prey could be more efficiently used. The maximum and minimum size that could be attained by a predator was constrained by the occupation of adjacent niches by other members of the 'hunting set'. These geographical variations in size work for several North American mustelid species. But there is a suggestion that other factors, besides interspecific competition, for example, habitat differences, might have a confounding effect by influencing both size and hunting set composition.

According to the 'reduction of competition' theories, both morphological types should benefit from their divergence. The greatest problem concerns the relative availability of prey to the two sexes. Wilson (1975) has demonstrated '... that while larger animals eat things which are unavailable to smaller competitors, the reverse is much less true'. This implies that there would be severe disadvantages in being the smaller of a pair; that is, the female. In practice, there is a consistent trend among mustelids for males to consume larger prey than females, and this is indeed the case for the mink (Birks & Dunstone 1985). So the female's food supply may be safeguarded by the male's preference for larger animals in the interests of his own efficiency.

The dimorphism in skull shape found in many mustelids, including mink, lends support to the avoidance of competition. The differences recorded are related to increases in the size of the canine and shearing (carnassial) teeth, or result from changes in skull shape required to accommodate the greater development of the more powerful jaw muscles in the male. Such adaptations may be interpreted as reflecting the importance of larger prey types in the male's diet. Alternatively, the relatively larger size of the canine teeth in the male may be explained by inter-male aggression for access to females.

While all the above arguments may represent benefits resulting from dimorphism, it is much less clear how they exert selective pressure to favour divergence in size. The evolution of different hunting strategies, foraging habits and activity patterns could reduce competition between the sexes without the need to evolve a difference in body size. Furthermore, an increase in the size of the male could lead to interference competition since social dominance in mustelids seems closely correlated with body size.

The main problem with the 'avoidance of competition' theory is that in some species (e.g. otters and badgers) even where there is substantial dimorphism in size and skull shape, there is no difference in the diet. Generally these theories fail to explain why in mustelids it is the male who is consistently larger than the female.

In summary, avoidance of intersexual competition for food is not a likely force for the evolution of sexual dimorphism in mustelids. However, secondary advantages may result from the differential use of prey which would serve to augment other selective pressures favouring the trait.

Reproductive roles

We must now consider the influence of differing reproductive roles in selecting for large males and small females. Sexual dimorphism is most pronounced in polygynous or promiscuous mammals, that is, those species where the male gains access to more than one female during the breeding season. Examples may be found in quite disparate groups including ungulates such as red deer, and primates such as baboons. In all cases there is a requirement for large body size arising from the strong inter-male competition for access to females. In mink, sexual selection seems to favour large males who are more likely to win fights and thus gain more mating opportunities.

This increase in body size will bring about a number of other advantages:

(i) greater mobility and hence a greater opportunity for encountering females—this arises because larger animals are able to travel faster and more economically than smaller ones (Peters 1983),
(ii) the larger home range size required to provide adequate nutrition is likely to encompass a greater number of female ranges,
(iii) the advantage of subduing females during mating fights.

The magnitude of sexual dimorphism observed in mustelids has probably resulted from the combined effects of selective pressures acting separately upon each sex. Undoubtedly, during evolution the advantage gained by the large male has led to his size being maximized within environmental constraints. The question remains as to how selection favours smaller female size; that is, what pressures act on her to counteract the genetic drift to largeness she inherits from her father?

Smaller females could be selected for because of their reduced energy demands, particularly during the breeding season (Erlinge 1979, Moors 1980). The energetic cost of reproduction in mammals is high; for some species costs increase by a quarter during gestation and may double during lactation. There is no doubt that the cost of reproduction is substantially lower for a small female than for a hypothetical male-sized female.

The idea that smaller females can make a saving on their own energy requirements which can be channelled into reproduction is not universally accepted. Ralls & Harvey (1985) argue that the evolution of small females would only be favoured if there was no difference in hunting efficiency between the sexes. In this context, 'efficiency' refers to the time and energy spent per unit of food energy captured. Although they may have a resting energy requirement 1.5 times that of females, male mink forage for about the same amount of time as females. This increased efficiency seems to be achieved by taking a greater proportion of larger prey. Thus, her diminutive size does not reduce the workload for the female. However, because small prey are more abundant she can satisfy her dietary needs with a succession of shorter foraging bouts. The result is that she need be away from her litter for only short periods. Indeed, when she is hunting for herself and her litter it is likely that she will have an equal energy requirement to the male.

So far discussion has focused on the implications of size itself. Growth, and the determination of size, is of paramount importance to the success of the male mink's first and subsequent breeding seasons. In these first few months of life, including and following dispersal, the mink are subject to severe competition, and selective pressure is at its most intense. Adult size is a function of both the rate of growth and its duration. However, both these parameters are variable and interrelated, and under the influence of the environment and genetic potential. Under ideal conditions growth rate is maximized until final body size is reached and growth is irreversibly arrested by the onset of puberty. In adverse conditions growth is slower, but puberty is primarily under seasonal control and cannot be delayed. The slower its growth rate, the smaller the adult mink will be.

Despite the slower growth rate of the female, she reaches adult body size in a shorter time, probably before food supplies begin to diminish with the onset of winter. This is an advantage conferred by her smaller size. For the male, growth is a race against time. Laboratory studies indicate that adult size in males is much more dependent on the plane of nutrition than is female size. Limiting energy intake in the first few weeks after weaning can significantly affect eventual adult body size. Any delay will push the male's growing period further into the lean season, when food shortage may cause growth to be prematurely arrested. The larger males with their greater maintenance requirements are more affected than the females by a restricted access to food. In the wild, male mink continue to grow at least until December, and the extent of sexual dimorphism may be related to the availability of large prey during this period. Sexual dimorphism is likely to be greatest in areas where food is abundant and males can realize their full growth potential.

Female size is a compromise. On the one hand she needs to be large to offset the energy losses caused by a large surface area to volume ratio, and on the other she will be able to channel more of her energy intake into reproduction if she has only a small body to maintain. But probably her greatest advantage is that she is able to grow and mature without unnecessary nutritional stress. Males pay a price for being larger as they must continue to grow well into the winter, despite lowered availability of food and the greater energy required to withstand cold weather and the forthcoming rut.

CHAPTER 6

Food and Foraging

MUSTELID PREDATORY BEHAVIOUR

Predatory behaviour involves a series of stages, including searching for prey, localization, recognition, pursuit and capture. The subfamily Mustelinae comprises small carnivores which actively pursue their prey. This subfamily includes the minks and weasels which are more specialized in their prey requirements than are the skunks (subfamily Mephitinae), which tend to be scavengers, and the badgers (subfamily Melinae) which are largely omnivorous.

Mink hunting behaviour is characterized by the occurrence of fairly rigid behaviours; that is, they show a high degree of stereotypy—a characteristic of innate (inborn) behaviours. There has been considerable debate over whether the behaviour patterns shown by mustelids killing prey are learned or inherited. Eibl-Eibesfeldt (1956, 1963) considers that only the correct orientation of the killing bite has to be learned, and that other prey-catching movements are innate but mature during the boisterous social and object play shown by young mustelids.

The small mustelids are amongst the most predacious species of carnivore and for their size they are formidable killers. Adult lagomorphs (rabbits and hares) would appear to be the upper size limit of native prey taken by mink in Great Britain. In North America the muskrat is a common large prey item of about equivalent size.

The typical method of attack is to leap onto the back of the prey, clutching its body with the fore-limbs. Preliminary bites may be delivered to any part of the body until the prey is secured or cornered in a confined space. The killing bite is delivered to the back of the head (the occipital region of the skull), with death resulting from damage to the hindbrain and/or the spinal cord. This technique is relatively less specialized than the application of the killing bite in the weasel and stoat. In some hunters like mongooses (family Herpestidae), the fangs penetrate the orbit of the eye. Comparative studies of the ontogeny of predatory behaviour have been conducted on polecats, stoats and weasels by Wüstehube (1960) and Gossow (1970), but there has been no detailed study of the mink in this respect.

One of my post-graduate students, Gregory (1987) carried out some preliminary observations on the utility or biological value to the predator of a range of prey items. The biological value concerns not only the nutritional quality of the prey, but also its abundance relative to other prey types, and the energy expended in search, pursuit and capture. With larger prey items, the cost of handling, dismembering and chewing the carcass may also be significant. Furthermore, the amount of indigestible and often difficult to handle material, such as bone, fur and feather, make up varying proportions of the total, depending on the size of the prey.

Mink were presented with the carcasses of laboratory mice and rats, wild rabbits, domestic fowl (1 day, 4 weeks and 12 weeks old), and a fish—the perch (*Perca fluviatilis*). Each corpse was presented for 36 h, during which time the animal's response was regularly observed. All the prey were accepted by the mink with the exception of perch. He records

> ... within the first 15 minutes the mink showed some 'killing' actions. In males this generally took the form of a single bite at the base of the skull or through the cranium, sometimes for periods of up to one minute. Often such 'killing bites' were followed by a frenzied thrashing around or a spinning motion possibly aimed at tearing through the skin of the carcass.'

Small prey, mice and chicks were consumed in their entirety. Rats were eaten beginning with the tail and hindquarters, then working up through the abdomen, picking flesh and bone from the skin and leaving the head and pelt until last, if they were eaten at all. With the larger domestic fowl it was more usual for the mink to start with the head or breast muscles. Two techniques were used when eating rabbits. The carcass could be entered through the soft underbelly or beneath the base of the tail; or alternatively, by starting with the head, eating down the neck and into the thorax. No part of the carcass of any prey species was found to be unpalatable except the contents of the lower alimentary tract of the rabbit. Occasionally feet, wings and shreds of skin remained after feeding. Birks & Dunstone (1984) have reported similar remains of prey (rabbits and sea-birds) from the dens of feral mink in Scotland.

AQUATIC HUNTING

There are two likely reasons why mink may choose to enter water in search of prey. Either their terrestrial prey becomes more difficult to capture than aquatic prey, or there is an increase in the ease of exploitation of aquatic prey. It is envisaged that at certain times of the year aquatic prey may become more available due to migration, for example, salmonids and eels becoming trapped in pools because of summer drought or even a receding tide. Gerell (1967b) suggested that the winter peak of fish in the diet of Swedish mink was due to the increased vulnerability of these cold-blooded (poikilothermic) prey when low water temperatures slowed their escape reactions, allowing even the slow-swimming mink a reasonable chance of capturing them. Such seasonal fluctuations in the availability (i.e. abundance and ease of capture) of prey necessitates a flexible hunting strategy. The mink shows a generalist hunting strategy, and is capable of modifying its foraging behaviour to hunt efficiently for the most available prey. This may be manifested as the mink selectively searching particular areas of the habitat for specific prey types, for example, eels on the bottom of a muddy pond or rabbits in warrens. By contrast, far more specialization is shown by the otter, which preys almost entirely on fish, and by the weasel, which hunts small rodents.

The aquatic hunting behaviour of the mink was first described in detail by Poole & Dunstone (1976) in a series of laboratory-based observations and experiments using hand-raised, ranch-bred mink, hunting for a variety of fish prey in large tanks. Mink behaviour was cine-filmed and then scrutinized using frame by frame analysis.

The mink had to be given experience of water at an early age or they never became proficient swimmers when adult. It was never possible to take an adult mink straight from the fur farm and expect it to dive underwater to hunt prey. It was a considerable, but very enjoyable, task each year to train the new batch of hand-reared mink kits by encouraging them to swim. This was usually achieved by joining them in their laboratory tank for protracted periods of aquatic play! It was remarkable how quickly these endearing animals quickly became untrustworthy, biting at any object proffered to them.

One of the most important and striking observations we made was that fish were generally detected by the mink before they entered the water. This stratagem has also been observed in the wild; as lone ago as 1892, Herrick observed a female mink fishing from a stone in mid river by plunge-diving on her prey. More recently, Melquist *el al.* (1981) recorded that foraging mink

> ... would travel along the shore or floating logs extending into the stream, pausing frequently to peer into the water for potential food. Once the prey was detected, the mink would quickly dive into the water after it.

A typical sequence of the events involved during fish capture is shown in Fig. 6.1. This was drawn from cine-film exposed at 24 frames s^{-1}, and the whole sequence lasts approximately 10 s during which time the mink has dived into a tank of water, pursued and captured a fish and exited from the tank. In the laboratory, prey location from out of the water was frequently enhanced by the mink peering intently into the tank while retaining a tentative grip on land with their hind feet. The next phase usually involved head-dipping. Here the mink immersed the head and sometimes even the thorax region underwater while still holding on to the tank. Eventually the mink would sight a fish and slip noiselessly into the water often pushing off from the tank edge as it dived.

Sometimes it was not possible for the mink to detect fish prey by viewing them through the water surface either because this was disrupted by ripples or

Mustelid Predatory Behaviour 65

FIGURE 6.1 *Typical dive sequence. Source: based on photographs by Nigel Dunstone.*

when the fish were given access to refuges in which to hide. Under these conditions one of the mink's strategies was to immerse the head (head-dipping). This allows the mink to search underwater efficiently since less energy is expended than when swimming, and breathing is facilitated by bringing the head out of the water for short periods. Continued failure to detect prey resulted in the mink moving to another position or resorting to underwater search. Head-dipping probably allows for the detection of fish up to distances of 1 m in clear water. Fish rarely showed escape movements when being viewed in this manner. Having detected the fish from a terrestrial vantage point, the mink's optimum strategy is to dive directly onto its prey, pushing-off from the substrate of the pool or tank to provide extra velocity.

Dive-chases, such as those described above, appear to be the mink's preferred aquatic hunting strategy. However, 'search-chases' were also used. These I defined as pursuits preceded by an underwater search phase (Dunstone 1978). Underwater searching occurred in the laboratory tanks, but this frequently took place after the fish had been initially located from out of the water and then lost. If the mink failed to maintain visual contact with the fish during pursuit, it frequently returned to the surface, left the water and attempted to relocate the prey from an aerial vantage point.

The duration of unsuccessful dives provides a useful indication of the mink's diving capability and underwater endurance (Fig. 6.2). If prey is caught, the dive is terminated and the mink returned to land to eat its catch, but an unsuccessful dive may be extended until the animal's oxygen supply is almost exhausted. Even so, it seems that mink rarely choose to reach this limit. The longest underwater search I recorded was 30 s, but the mean duration of such dives was a mere 9.9 s. Oxygen limitation ultimately causes a search or pursuit manoeuvre to be discontinued. This leads to a negative correlation between the search duration and the pursuit duration of a given hunt; that is, as the search phase is prolonged, the maximum duration of any pursuit then initiated if the mink finds a fish is steadily reduced. Fig. 6.3 shows that search-chases are on average 3 s shorter than dive-chases. This clearly demonstrates the effect of the oxygen constraint on the pursuit phase of a search-chase, where the mink did not have the opportunity to fill its lungs before pursuing the prey.

We would thus expect a trade-off between the benefits of continuing search and the cost of losing the prey on oxygen exhaustion during pursuit. This may explain why the mink decides to terminate most of its dives before that physiological limit is reached. It prematurely abandons an unsuccessful search in favour of returning to the surface, replenishing its air supply and possibly re-locating a fish from out of the water.

High exertion levels during diving further exacerbates the problem of suspended breathing. The mink's relatively inefficient underwater propulsion requires a high power output (see Chapter 5). In addition, metabolic heat production must be elevated to balance the high rates of heat loss when immersed in water.

Using a multivariate statistical analysis of dive durations, Dunstone &

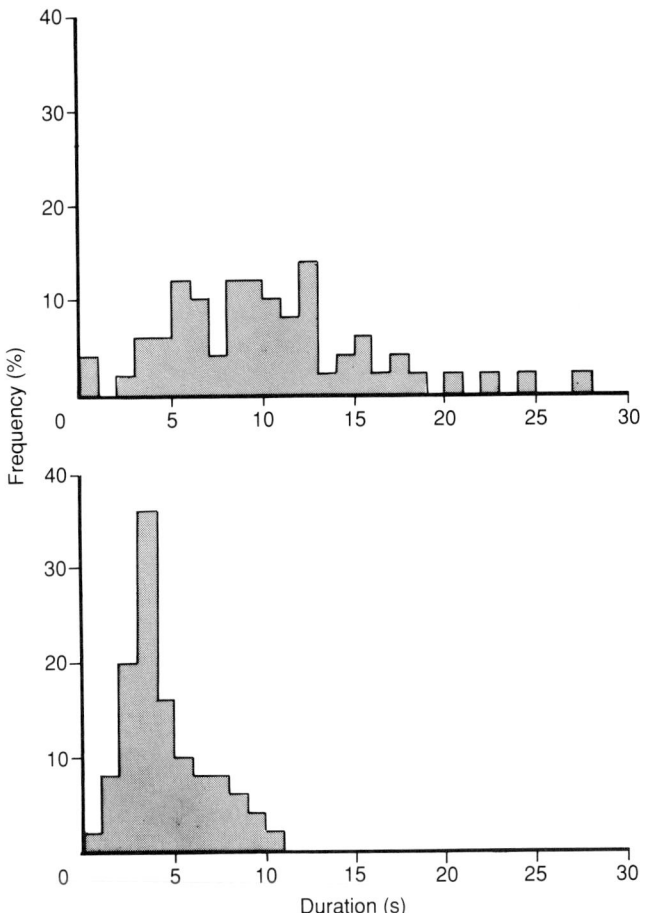

FIGURE 6.2 *Frequency distribution of the durations of unsuccessful (top) and successful (bottom) dives recorded in the laboratory. Source: Dunstone (1978).*

O'Connor (1979a,b) showed that only 23% of the variation in the observed frequency and duration of dives could be explained by the availability of a limited oxygen supply, and that the greatest part (51%) of the variability in diving behaviour was due to factors related to the optimization of foraging activity.

The mink seemed able to retain a spatial memory of the topography of the tank using information provided by the position of refuges for the prey and other features, since successive searches did not tend to re-visit areas of the tank already unsuccessfully explored. The provision of refuges caused a redirection of the mink's search effort to these structures where fish could frequently be found. The mink responds to a capture in or near a refuge by

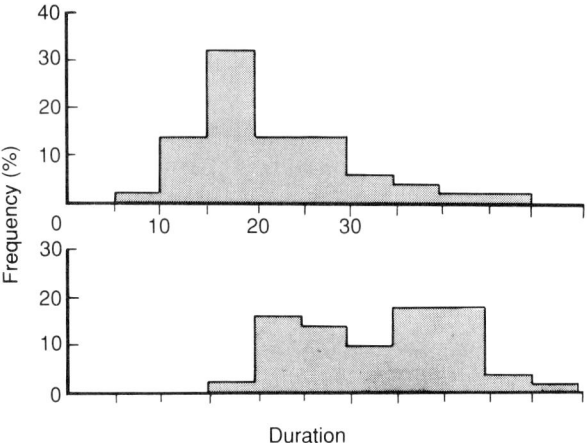

FIGURE 6.3 *Frequency distribution of the durations of search-chases (top) and dive-chases (bottom) recorded in the laboratory. Source: Dunstone & O'Connor (1979a).*

increased investigation of that area on subsequent dives. Thus, the mink quickly learns the places where it is likely to encounter prey rather than searching open water. Presumably this also occurs in nature, with the mink selectively searching areas of its habitat where it has previously encountered prey. When there were no refuges available for the fish to hide in and the fish were restricted to open water the mink's hunting efficiency, measured by the proportion of successful dives, increased with the increasing number of fish present.

Experiments conducted with varying numbers of fish prey (minnow, *Phoxinus phoxinus*) present have demonstrated that their shoaling behaviour markedly decreases the mink's hunting efficiency. When hunting the shoal of minnows the mink dived into the shoal, thus dispersing it, and then attempted to isolate individual fish from the group. However, 'explosion' of the shoal around the diving mink frequently resulted in the predator becoming confused and failing to capture a fish (Poole & Dunstone 1976).

The longer the pursuit, the greater the probability of the mink capturing the fish. Although most fish species were capable of 'burst-swimming' faster than a pursuing mink, it seems that the mink are able to fatigue their prey, and their sinuous necks provide additional reach to grab at the fish. Fish were not attacked in the same way as mammalian prey, but grasped wherever the mink made contact. Frequently fish were removed from the pool with a tenuous grip on a fin, only to slip back in and subsequently escape.

Much work has been done on the diving behaviour of another aquatic member of the family Mustelidae, the otter. Since the habits of mink and otter overlap it is interesting to compare the dive times for the two species. Kruuk & Hewson (1978) recorded mean dive durations of otters foraging in the sea off the coast of Scotland of 16 s for successful dives and 25 s for unsuccessful dives, with a maximum time spent underwater of 35 s. Observations made by Conroy & Jenkins (1986) of otters fishing in freshwater lochs gave values of 13.1 s for successful dives and 12.7 s for unsuccessful dives. It would appear that, in fresh water at least, otters and mink forage underwater for similar durations despite the greater size and swimming prowess of the former.

In addition to the constraints noted above, environmental conditions may serve to further reduce the mink's foraging efficiency. Illumination level and water clarity have been shown to affect directly the mink's underwater vision. By manipulating features within the tank (e.g. water depth, current flow, habitat complexity and prey density) it is possible to investigate the effect these variables have on the mink's diving ability.

Apart from ethical problems, one of the major difficulties encountered in the study of predation of live fish is the highly variable and unpredictable behaviour of the prey as it attempts to escape from the predator. The most thorough analysis of diving behaviour of the mink has been conducted by another of my post-graduate students, Sharon Davies, who, in a series of exhaustive laboratory tests, observed that the mink could vary their dive endurance to cope with a variety of environmental conditions including water depth, habitat complexity and current flow. Davies (1988) avoided the problems inherent in using live prey by training mink to open wooden boxes—some of which contained pieces of fish flesh—distributed on the floor of a large 7×4 m tank which held water to a depth of up to 1.65 m (see Fig. 6.4). Accurate dive durations and swimming distances were obtained by video-taping the dives made by mink. Figure 6.4 also shows just one of the thousands of dive paths she recorded from mink as they searched for 'prey' in the tank.

As the water depth increased from 0.3 to 1.65 m, the dive rate decreased (Fig. 6.5). This was partly due to the mink taking a longer recovery period between dives, with longer dives requiring a longer inter-dive interval. The duration of dives also increased at greater depths by a factor larger than the additional time required to get to the foraging area on the bottom of the pool. Generally, at the shallower depths the mink made more shorter dives, whereas when the water was deep, they made fewer, longer dives.

An important feature of the riverine and coastal habitats frequented by mink is the presence of variable strength water currents. At the very least this would be expected to influence the costs of locomotion during hunting, either by increasing the energy requirement when swimming against a current or minimizing it if travelling downstream. Given the poor swimming ability and dive endurance of the mink, it might be expected that they would not forage in such areas. Davies (1988) employed two types of current in her experiments. A deep current caused a movement of the water throughout the tank at a maximum speed of 0.8 m s^{-1}. Although this is a fairly weak current, it is representative of that found in areas where mink have been observed hunting. The alternative experimental condition used a surface current flowing at approximately 0.6 m s^{-1}. The effect of either type of current flow on the

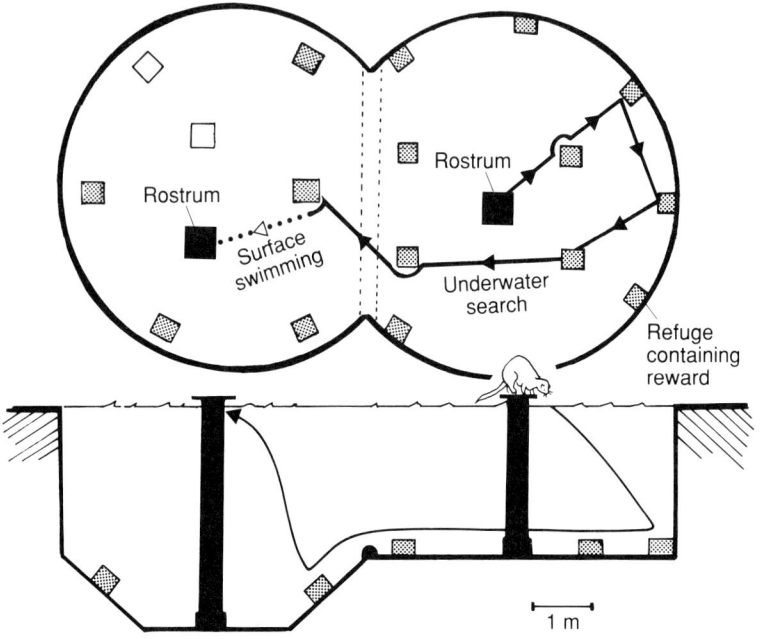

FIGURE 6.4 Typical search path followed by mink while foraging underwater. Source: based on Davies (1988).

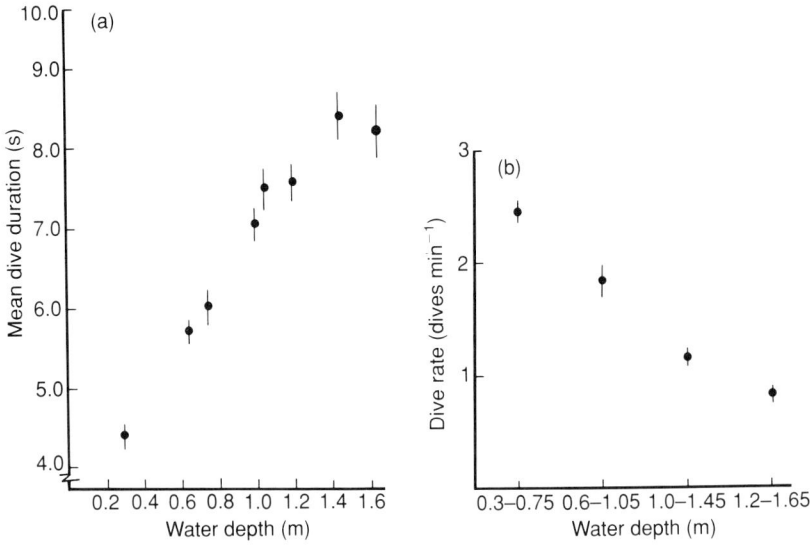

FIGURE 6.5 Diving endurance of the mink. The effect of water depth: (a) dive duration, (b) dive rate. Source: Davies (1988).

diving endurance of the mink was negligible. Dive durations and dive rate were unaffected. There was an indication from those experiments using a surface current that the disruption to the water surface in the form of small waves inhibited foraging activity to some extent, possibly by interfering with the mink's preferred strategy of sighting prey (or in this case baited food boxes) from out of the water.

In a theoretical study, Kramer (1988) applied optimal foraging principles to the economic use of air by diving mammals. He predicted that with increasing depth, both time on surface and dive duration would increase. This prediction was supported by the findings of Davies (1988) with mink. Kramer (1988) also predicted that as the distance to feeding site increased, the amount of stored oxygen should be increased to allow a greater time away from the surface. I have frequently observed mink panting, presumably to load their lungs and blood with oxygen prior to diving.

Observations of hunting in the wild
Given the intensity of effort that has been put into mink research world-wide, there are few published observations of predatory behaviour in this species in the wild. Mech (1965) reports his observation of

> ... a caged mink [which] tackled a house rat not much smaller than himself. With a predator's swiftness and agility, the mink lunged and locked his jaws in the loose skin of the back of his victims neck, a rather precarious hold. The rat almost escaped, but the mink flopped over on his side and dug in with claws and teeth. In a few moments it was all over.

Some years ago a research colleague reported a mink's attempt to subdue a hare whilst riding around on its back (J. Birks pers. comm.). The hare escaped. Observations made during his post-graduate research led Ireland (1990) to suggest that rabbits are commonly hunted down their burrows. However, he occasionally observed struggles between mink and rabbits above ground, but on these occasions the mink was never seen to kill its prey:

> ... the rabbit broke free of the mink's grip, though one [mink] was observed pursuing, and keeping up with, a bolting rabbit in an open field for almost 50 metres before the rabbit escaped.

Direct observation of these elusive creatures is difficult. Melquist *et al.* (1981) radio-tracked 26 individual mink for a total of 889 h, but animals were actually observed foraging for a mere 2.1 h within that period! While working on coastal mink in Scotland, Mark Ireland, Johnny Birks and myself actually observed mink for 5 h out of 1500 h spent radio-tracking. Only rarely were observations made of radio-collared animals foraging in rock pools on the shore, and swimming and diving in the sea, even though prey from these habitats comprise a significant proportion of the diet (Dunstone & Birks 1987).

In British Colombia, Hatler (1976) had considerably more success in watching the antics of hunting mink. The mink he observed during 136 hunting bouts used three main strategies:

Bird-dogging: this was used mainly on eel-grass flats and involved moving along, in either a straight line or while zig-zagging, with the nose held close to the ground. The head is moved from side to side as the mink tries to pick up the scent of its prey. Crabs were commonly located in this way. The mink deals with them, after excavation from weed or sand, by flipping them over on their backs and biting through the shell. Stranded fish were also located in this way, although the sound of their movement might have provided an additional cue.

Poking: whereby the mink sticks its head into crevices and under boulders to obtain fish and small invertebrates stranded by the tide. This can even involve burrowing beneath mounds of algae.

Diving: most dives were into water 2–3 m deep, although on one occasion a mink was observed to forage successfully for a bottom-dwelling crab, in an area subsequently measured to be 7.4 m deep.

Table 6.1 shows how the use of these various strategies varied with the nature of the shore. The duration of dives was longer at high tide compared to low tide, presumably indicating the effect of increased depth in a similar manner to my laboratory experiments. Figure 6.6 presents the durations of the 264 dives that were observed and indicates what proportion were successful. Most of the dives were surface dives. Hunting underwater seemed to be similar to hunting on land with the mink exploring underneath likely boulders for its prey. That the mink occasionally brings dead crabs or their

TABLE 6.1 *Hunting methods used by mink in various coastal habitats, on Vancouver Island, British Columbia.*

Habitat	Bird-dogging		Poking		Diving		Mixed method	
	No.	%	No.	%	No.	%	No.	%
Boulder beach ($n = 106$)	3	2.8	2	67.9	31	29.2	20	23.2
Rockweed shore ($n = 61$)	2	3.2	22	36.1	37	60.1	8	15.1
Eelgrass flats ($n = 61$)	36	59.0	5	8.2	20	32.8	18	42.9
Small particle beach and estuary ($n = 8$)	1	12.5	3	37.5	4	50.0	2	33.3

Source: Hatler (1976).

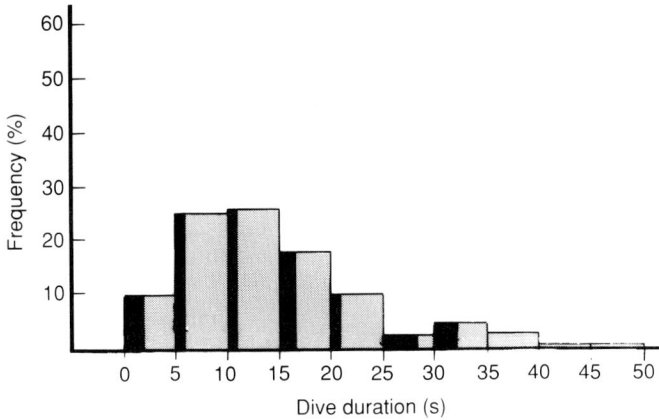

FIGURE 6.6 *Frequency distribution of the dive durations recorded from wild mink hunting in a coastal habitat. The shaded portion indicates the proportion of the dives that were successful. Source: Hatler (1976).*

cast-off skins to shore indicates that they are probably using touch or vision to locate 'prey', in particular, shape perception for the recognition of potential prey. It would appear that movement was not a relevant cue.

Hatler (1976) also documented a novel technique of fishing that he referred to as 'herding'. In this instance the mink swims parallel to the water's edge but slightly offshore and chases fish onto the beach, where they are caught before they can return to the water.

Captured prey were taken to the shore and either eaten immediately or cached; 66% of crabs were cached, while there was an equal likelihood of the mink eating or storing any captured fish.

SURPLUS KILLING AND THE CACHING OF FOOD

Many carnivores exhibit the phenomenon of surplus killing under conditions where the prey are encountered at abnormally high density and cannot escape from the predator. The phenomenon is not unique to mink, but tends to receive considerable publicity when it does occur.

Surplus killing of domestic stock may represent a panic response when the predator is surrounded by excited prey such as hens in a chicken shed. Such incidents can be avoided by good husbandry techniques (see Chapter 11).

Surplus killing incidents are unusual in the wild, and when they do occur they often represent an attempt by the predator to cache prey. The habit of caching food has also been occasionally reported in laboratory studies of mink (Sinclair *et al.* 1962, Bleavins & Aulerich 1982). This trait has been interpreted as an adaptive behaviour conferring an advantage upon opportunistic predators which exploit food resources of varying abundance and availability.

The circumstances surrounding food caching behaviour in carnivores vary; the most common reason is to hide the food from the attention of scavengers, particularly when it is too large to be eaten in one meal. For smaller carnivores, caching may follow opportunistic surplus killing. Various prey types have been discovered, including the remnants of pinioned waterfowl, animals from an experimentally transplanted cotton-rat population and overwintering (torpid) frogs. Food storage may buffer the mink against periods of poor weather when it cannot hunt effectively.

Caching may free more time for foraging. This may be particularly important in situations where foraging time is restricted, for example, at low tide when fishing has to be completed before the tide returns. After obtaining sufficient food, the mink can return to its cache to eat at leisure.

Apart from these spectacular observations, few people have had the opportunity to study the foraging behaviour of mink in detail, and even then it was frequently not possible to see what the mink had captured. It is because of this paucity of observations that scientists have had to resort to other, more clandestine, means of investigating the mink's diet.

DIET

Ideally we would wish to observe predatory events as they occur so that the type and size of the prey, and the relative costs and benefits of the meal can be accurately assessed. Only rarely is this possible for mammals and especially difficult for mink under most circumstances, even when it is the subject of intensive field investigation. Hence, we have to resort to deducing the animal's diet by examining the remains of meals, by analysing gut contents or faecal remains. Each method has attendant advantages and disadvantages as to the quality and reliability of the information that can be inferred.

ANALYSIS OF FOOD REMAINS FOUND AT MINK DENS AND FOOD CACHES

A few studies have looked at prey remains found in mink dens. The advantage of this technique is that accurate identification, often to species level, can be made, and information on the age (adult or juvenile) of the prey may be available. The information obtained on overall diet of mink will be incomplete, however, since not all the dens of the mink may be known, small items of prey may be eaten at the site of capture, and only larger items may be brought back to dens. Yeager (1943) investigating reports of hoarding at mink dens in North America, found the main prey items to be muskrats, but waterfowl, and in one case seven fish, were also found. In one mink den located in a hollow ash log, 13 muskrats, two mallard ducks and a coot were found. (The culprit mink was subsequently caught, the muskrats skinned, and the 14 pelts sold for $US37!). Sargeant *et al.* (1973) found the remains of a radio-transmitter equipped blue-winged teal (*Anas discors*), along with five other adult birds, under a pile of branches close to an experimental pond affected by mink predation. A search of 16 other dens on the semi-permanent marsh system revealed a large variety of avian prey (Table 6.2), many of these birds having been just released from rearing pens. Eberhardt & Sargeant (1977) examined the remains of 203 partially consumed prey found at 68 muskrat burrows that had been used as dens by mink along marsh shorelines; 78% of the prey were birds, with waterfowl the major component (Table 6.3).

TABLE 6.2 *Composition of food remains at 16 mink dens on a semi-permanent marsh in eastern North Dakota.*

Prey	Frequency
Waterfowl	
Wood duck duckling	21
Unidentified wild duckling	2
Blue-winged teal	3
Ruddy duck	2
Gadwall	1
Mallard	1
American coot	15
Grebe	5
Duck egg	8
Other birds	
Red-winged blackbird	5
California gull	1
Hungarian partridge	1
Unidentified small bird	2
Miscellaneous	
Thirteen-lined ground squirrel (*Spermophilus tridecemlineatus*)	1
Tiger salamander (*Amblystoma tigrinum*)	1

Source: Sargeant *et al.* (1973).

In Wales, Hill (1964) reported a cache of 21 trout (20 cm long) found in a river bank, which he attributed to mink predation. Birks & Dunstone (1984) recovered 96 items from coastal mink dens in Scotland (see Table 6.4); 62.5% were mammal remains, mostly rabbit (*Oryctolagus cuniculus*), and the remainder were birds. Of these, 21 items (21.9%) were sea-birds, including four waders. A pheasant (*Phasianus colchicus*) and a grey partridge (*Perdix perdix*) were also found. The remainder were crows (*Corvus* sp.) and pipits (*Anthus* sp.).

In Russia, Danilov & Tumanov (1976) found frogs, burbot (*Lota lota*) and a mallard duckling in the dens of three breeding female European mink.

TABLE 6.3 *Species and sex composition of 94 adult ducks found at mink dens in east-central North Dakota.*

Prey species	Number found at dens			
	Males	Females	Unknown	Total
Mallard	6	3	8	17
Gadwall				
(*Anas strepera*)	3	2	3	8
Pintail	2	1	3	6
Green-winged teal				
(*Anas crecca*)	1	1	0	2
Blue-winged teal	18	6	6	30
American wigeon	0	0	1	1
Northern shoveler				
(*Anas clypeata*)	1	2	4	7
Total dabblers	31	15	25	71
Redhead				
(*Aythya americana*)	0	4	0	4
Canvasback				
(*Aythya valisineria*)	0	2	0	2
Lesser scaup				
(*Aythya affinis*)	1	1	0	2
Ruddy duck	3	7	5	15
Total divers	4	14	5	23
Total	35	29	30	94

Source: Eberhardt & Sargeant (1977).

ANALYSIS OF GUT AND FAECAL CONTENTS

The investigation of mink diet has mainly involved the analysis of mink scat (faecal remains), or on gut contents. The advantages of scat analysis, apart from their relative ease of collection, is that the diet of local populations can be studied over long periods and information obtained on variations in diet attributable to season, sex, individual differences and so on. The main disadvantage is that scats contain only those parts of the food intake which cannot be digested (bones, hair, feathers, scales). Thus, foods consisting mainly of soft parts will be under-represented in the sample giving misleading results. This can arise if, for example, the predator removes only the flesh from a large prey animal it has killed. A further complication arises because different species and sizes of prey have different proportions of parts which pass through the gut undigested. This may lead to items containing a high proportion of undigestible remains (e.g. Crustacea) being over-represented in the dietary analysis. To try to overcome this, scientists have developed several methods of presenting the results of scat analysis, all aiming to produce the best fit between actual prey intake and identifiable items in the scat.

Various methods of presenting the results have been used. Commonly, percentage frequency (i.e. the percentage of scats containing a particular

TABLE 6.4 *The occurrence of prey remains recovered from mink dens on the Galloway coast, Scotland.*

	Frequency	Percentage
Mammalian prey		
Rabbit (*Oryctolagus cuniculus*)	50	52.1
Brown hare (*Lepus europaeus*)	6	6.3
Brown rat (*Rattus norvegicus*)	2	2.1
Hedgehog (*Erinaceus europaeus*)	1	1.0
Lamb tail (*Ovis aries*)	1	1.0
Sub-total	60	62.5
Avian prey		
Herring gull (adult) (*Larus argentatus*)	6	6.3
Herring gull (chick) (*Larus argentatus*)	9	9.4
Black-headed gull (*Larus ridibundus*)	2	2.1
Common/Arctic tern (*Saterna* sp.)	1	1.0
Guillemot (*Uria aalge*)	2	2.1
Razorbill (*Alca torda*)	1	1.0
Woodcock (*Scolopax rusticola*)	1	1.0
Lapwing (*Vanellus vanellus*)	2	2.1
Oystercatcher (*Haematopus ostralegus*)	1	1.0
Pheasant (*Phasianus colchicus*)	1	1.0
Grey partridge (*Perdix perdix*)	1	1.0
Corvus sp.	2	2.1
Columba sp.	2	2.1
Anthus sp.	5	5.2
Sub-total	36	37.4
Total	96	

Source: Birks & Dunstone (1984).

item), or relative frequency (i.e. the number of occurrences of a prey item expressed as a percentage of the total number of occurrences of all items in the sample) are used. Wise *et al.* (1981) have developed a bulk percentage measure which aims to give information on the relative importance of the different prey types found in the diet.

Several workers have attempted to calculate correction factors based on feeding trials with captive animals (e.g. Akande 1972, Fairley & Ward, 1987 for mink; Lockie 1959, for foxes, *Vulpes vulpes*) in an attempt to get around these problems, typically with little success. For example, Fairley & Ward (1987) supplied a wide range of the mink's naturally occurring mammal, bird and fish prey to captive mink. However, the correction factors obtained were of limited use due to the wide variation in the values, both between different species of prey, and also between different sizes of prey and predator involved. Since we rarely know the sizes of prey taken or that of the predator eating them, the use of correction factors has been very limited. Hence, most authors continue to use an uncorrected percentage frequency or the relative frequency of occurrence.

Analysis of gut contents would seem to give a better picture of the recent diet of the individual since larger fragments or even intact prey may be obtained. Unfortunately this method has the obvious limitation that a large proportion of the mink population may have to be sacrificed to obtain the result, and hence does not allow for continuity of study. Large samples may be available for parts of the year only; for example, Hamilton (1959) was able to obtain mink carcasses from trappers during the open season (autumn and winter), and had to rely on scat analysis to determine the spring and the summer diet.

Despite the limitations of the methods and the variety of analyses performed, it is possible to draw together the results from the large number of studies that have been conducted. Prey items have been grouped into major

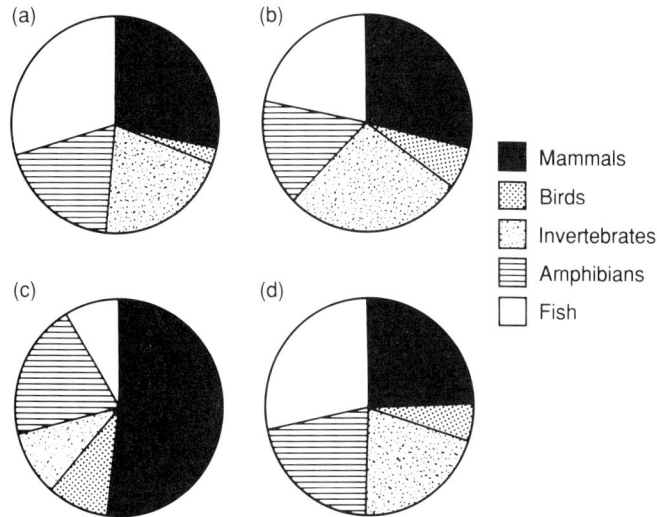

FIGURE 6.7 *Winter and summer diet of the mink in North America. (a) New York state, winter, (b) New York state, summer, (c) Michigan, winter, (d) Missouri, all year. Source: based on (a & b) Hamilton (1959), (c) Sealander (1943), (d) Korschgen (1958).*

categories. Some investigators have recorded the presence of vegetation in mink guts (Korschgen 1958, Hamilton 1959) and scats (Day & Linn 1972), but others believe vegetation is ingested incidentally (e.g. Wise *et al.* 1981). The data have been recalculated to give relative frequency of occurrence to allow a comparison between studies. Figure 6.7 summarizes the information obtained from scat and gut content analyses of wild mink in their native North America. Data for mink studied in those countries where there are only feral populations are given in Fig. 6.8(a) and (b).

The mink is clearly an opportunistic predator, taking a wide variety of

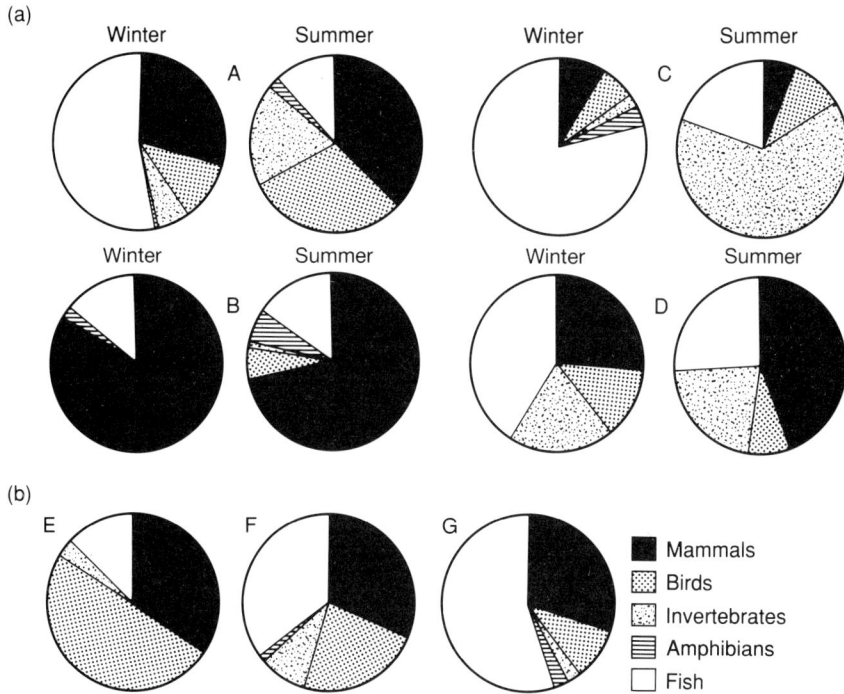

FIGURE 6.8 *(a) Winter and summer diet of the mink in a variety of habitats in Europe; A bog; B lake; C river; D coast. Source: A,B,C, Gerell (1969); D, Dunstone and Birks (1987). (b) Diet of mink in southwest England: E, eutrophic lake, F, oligotrophic stream, G, eutrophic stream. Source: Chanin & Linn (1980).*

mammal, fish, bird and invertebrate prey. Many authors, notably, Gerell (1968) and Dunstone & Birks (1987) have noted seasonal variations in dietary composition, and that diet also varies for each locality of scat collection.

The catholic diet of the mink makes possible only the broadest comparisons across various countries. Generally, the diet in all countries consists of a mixture of prey types in varying proportions dependent on what is locally available. Only occasionally have researchers managed to assess the availability of prey in the area from which scats have been collected. Where this has been achieved, it has been suggested that mink appear to take prey in relation to their availability, that is, local abundance and vulnerability. The overall picture obtained from these studies is that mink diet is very similar in general composition to that in North America, although of course the species taken frequently differ.

In Europe, fish often seem to be the most common prey, whereas in North America mammals are reported to be more frequent prey items. Birds, especially waterfowl, seem to be of greater importance in the diet of feral mink in Europe than in their native America. A commonly taken mammal in North America is the muskrat. However, since this species is not widely distributed in Europe, other large mammals, particularly rabbits, are used as prey.

PREY GROUPS AND SPECIES TAKEN

Mammalian prey in North America

The relative frequency with which mammal prey is found in the scats of mink can very from 3.0% in a coastal habitat (Hatler 1976) to 54.7% on rivers and marshes (Sealander 1943).

In North America the largest and probably the most important mammal prey is the muskrat. The mink appears to be an important winter predator on this species (Wilson 1954, Errington 1954, 1961). Muskrats, particularly young animals, are vulnerable to mink predation during low-water winters. At other times mink can subsist largely on muskrat carrion.

Mink and Muskrats The exploitation of one valuable fur-bearer, the muskrat, by another, the mink, to the detriment of commercial hunting was a considerable stimulus for a lifetime of research conducted by Errington, who contributed greatly to the early study of mink ecology. The muskrat can be a formidable prey for a mink (Errington 1954):

> No one should assume that minks, professional fighters though they are, care nothing about wounds nor that minks experienced in the ways of formidable muskrats feel impelled to risk severe bites just for the sake of a meal if they can feed upon something else more cheaply. . . . if a mink overtakes a panicky muskrat trying to run away on the ice, killing may be easy. If a big muskrat goes after a mink with mayhem in mind and carving tools in front, the mink may not care to accept the sporting challenge.

Errington has documented the interactions between these two animals on a year-round basis in stream and marsh areas in Iowa since 1943. Several categories of mink predation upon muskrats were distinguished. During the breeding season, adult muskrats that possessed 'regular' home ranges were virtually safe from predation by mink, as long as their habitats remained in good condition. However, if the quality of that habitat declined (e.g. through drought), the population could suffer severe mink predation. Those adult muskrats which did not possess a territory at this time were forced to become more active on land and less able to escape. Such animals included those surplus individuals in the population that frequently become injured during intraspecific fights with territory owning individuals. Mink preyed heavily upon young muskrats whenever there was over-population and the surplus were forced to 'adopt a hazardous way of life in consequence of attacks from other muskrats when forced ashore from the crowded wetter parts of the marshes' (Errington 1954).

Following breeding, many surplus young muskrats were removed from the population by mink; then predation fell to negligible levels in late summer and autumn, unless inclement conditions brought about crises which increased the rodents' vulnerability. From mid winter through early spring muskrats that came out of their burrows onto the surface of snow or ice to seek food were particularly vulnerable, as were those animals that ventured early from their burrows with the arrival of the new breeding season.

Generally then it would appear that if anything goes wrong with the everyday fortunes of the muskrats, the mink were usually there to seize the opportunity. One example of this involved (Errington 1954):

> ... a storm blew ashore a group of lodges that had been insecurely anchored in dead cattails, thus precipitating fighting between the displaced muskrats and those defending their shore zone territories and forcing many animals to wander along the shore in the vicinity of the den occupied by the mother mink and her litter. The mother mink had not been known to prey upon muskrats earlier in the summer, but, during or immediately after the storm, she was known to have brought 16 young muskrats to her den.

Of considerable importance is the mink's use of muskrat carrion. On one occasion a density of 20 muskrat corpses per acre were available to mink. This bountiful supply of food resulted from a haemorrhagic disease to which the muskrats were susceptible. 'As long as muskrats kept dying the local minks took advantage of them, though overlooking two dead for every one that they found' (Errington 1954).

There are plenty of places in North America where free-living mink maintain themselves adequately in the absence of muskrats, so the muskrat should not be thought of as being indispensible to the mink's way of life. Nevertheless, the mink is well suited to the habitats where muskrats live. This brings us to the idea that this coexistence may result, at least partly, from the engineering activities of the muskrats. The muskrats may, inadvertently, provide denning facilities for the mink, and even passageways to live food supplies under the ice.

In areas where muskrats were not available (e.g. in Missouri; Korschgen 1958), the cottontail rabbit (*Sylvilagus floridanus*) is commonly taken instead. Generally of less importance is the assemblage of small mammal prey. Species taken include the prairie vole (*Microtus ochrogaster*), meadow vole (*Microtus pennsylvanicus*), red-backed voles (*Clethrionomys gapperi*), deer mouse, white-footed mouse (*Peromyscus* spp.) and ground squirrels (*Spermophilus* sp.). Only rarely (Korschgen 1958) do these species contribute a significant proportion of the diet. Unusual mammalian food items such as bats (*Myotis* spp.), taken in a Kentucky cave, occasionally turn up in the diet (Goodpaster & Hoffmeister 1950).

Mammalian prey taken in Europe

In some areas of Scotland, rabbits were the predominant mammal taken (Jenkins & Harper 1980, Dunstone & Birks 1987). In other parts of Scotland, woodmice (*Apodemus sylvaticus*), bank voles (*Clethrionomys glareolus*), and field voles (*Microtus agrestis*) made up the bulk of the diet (Akande 1972, Cuthbert 1979).

On the River Teign in Devon, Chanin & Linn (1980) found mink took a wide variety of mammalian prey, including species which were comparatively scarce such as dormouse (*Muscardinus avellanarius*) and the harvest mouse (*Micromys minutus*). Results from Slapton Ley, a lake habitat, showed mink to concentrate their predation on rabbits, while on the River Frome, field voles and rats (*Rattus rattus*) were frequently eaten. Also working on the diet of mink at Slapton Ley, Wise *et al.* (1981) have shown woodmice and voles to be seasonally important. These studies showed that mink occasionally take insectivores (shrews, *Sorex araneus* and *Sorex minutus*, and moles, *Talpa europaea*), albeit at low frequency. This is somewhat surprising since many carnivores are reputed to find them distasteful. In Sweden, in a series of extensive studies conducted by Gerell (1967b), field voles, rats and water voles were commonly taken.

At least four studies have recorded the presence of mustelid (probably weasel) in mink scats (Hamilton 1959, Cuthbert 1959, Chanin & Linn 1980, Chanin 1981), but the circumstances under which this was taken are unknown.

BIRD PREY

Identification of bird remains from scats is not possible to species level. Only when remains of kills are found can species be accurately identified. The groups most commonly predated are water inhabiting birds (ducks, moorhens, coots) and waders.

Bird prey taken in North America

If questioned concerning the impact of mink on native wildlife, many country people will immediately state that since mink invaded the local river or lake, numbers of breeding waterfowl will have declined dramatically. Some research studies have also suggested that mink can have a considerable local effect on ground-nesting avian prey. Eberhardt & Sargeant (1977) investigated this interaction on a population of prairie marsh-living mink in North Dakota, using carcass remains at dens and scat analysis (see Table 6.3). Individual mink families were studied. One such family ate three of the estimated 36 adult coots and 76 of their young. They estimated that the adult coots could have produced 144 young, and that only 26 juveniles were seen at the end of the breeding season, the mink having accounted for 8% of the adult and 52% of the juvenile population. This same mink family simultaneously accounted for 11 juvenile ducks, but no information was collected on the number of ducklings consumed. This situation may have been exceptional, since, in the following year, there was only a small decline in waterfowl breeding success even though a mink and her kits were present on the marsh. The vulnerability of coot chicks may be due to their very vocal behaviour immediately after hatching, which attracts the attentions of the mink.

Juvenile ducks seem less susceptible to mink predation than juvenile coots. Parent ducks have been observed defending their broods against marauding mink, whereas this does not appear to happen in coots. It is equally possible that the predation on coots buffered, to some extent, potential predation upon juvenile ducks. Diving ducks were found to be more at risk than dabbling species. Of the dabblers, only 33% of those taken were adult females compared with 78% of the diving duck species. The differing sex ratios of the two types of duck taken may be explained by their divergent nesting habits. Diving ducks habitually nest close to water and are more likely to be encountered by the mink.

Predation on birds has been reported in all studies, with the highest levels directed towards waterfowl and game birds. Eberhardt & Sargeant (1977) found predation on waterfowl to be particularly intense during the breeding season. Birds accounted for 78% of the prey taken, and of these, 86% were ducks, commonly blue-winged teal or mallard, and American coots (*Fulica americana*). Pied billed grebes (*Podilymbus podiceps*) were also occasionally taken. Comparing their relative availability, pintails (*Anas acuta*) were taken less frequently and ruddy ducks (*Oxyura jamaicensis*) more frequently than expected. Just under half of the ducks taken were juveniles.

Songbirds including the flicker (*Colaptes auratus*), cardinal (*Cardinalis* spp.), bluebird (*Sialia* spp.), red-winged blackbird (*Agelaius phoeniceus*), yellow-headed blackbird (*Xanthocephalus xanthocephalus*) have also been recorded in mink diets (Sealander 1943, Korschgen 1958, Arnold & Fritzell 1987), although all were at very low frequency.

Bird prey taken in Europe

Waterfowl can form a significant portion of the mink diet. It is thought that these are taken in the shallow areas near lake and river banks or on the shore while roosting (Wise *et al.* 1981). Occasional anecdotal observations have been made of mink snatching swimming birds from under the water. Chanin & Linn (1980) working in Devon found that mink took bird prey two to three times more frequently on the River Frome and Slapton Ley (a lake) than on the River Teign. They related this to the fact that the bulk of the avian prey on the lake (mainly coots, moorhens and ducks) are generally found close to or on water, where their poor flying abilities may make them more vulnerable to mink predation. Chanin (1981) suggests that coots and moorhens are more easily killed by mink than are the larger ducks since they spend more time in reed beds where they are accessible to ambush by mink. On the River Teign, fewer bird species in total were taken since those in abundance were the more terrestrial species, and hence less accessible types, including passerines and woodpigeons. Gerell (1967b) also notes waterfowl, particularly mallards and coots to be commonly taken in Sweden.

Different species of bird prey are available to coast-dwelling mink. Not surprisingly, their diet is found to contain sea-birds and waders (Charadriiformes) in addition to those species also noted on inland waterways (Birks & Dunstone 1985, Dunstone & Birks 1987). The wide range of bird prey

FIGURE 6.9 *Herring gull carcass attributed to mink predation.*

species found at a coastal mink den in Scotland by Birks & Dunstone (1984) (see Table 6.4) included many species of sea-birds. Some species were available locally, such as the herring gull (Fig. 6.9), but others were more likely to be have been obtained as carrion after winter storms. Sea-birds, principally terns *Sterna paradisaea*, have also been reported in the diet of mink from Iceland (Skirnisson 1980).

AQUATIC PREY

Fish

Unfortunately, when the diet of mink is investigated by means of scat analysis, information on the species of fish eaten may be poor because the diagnostic vertebrae, and scales may be damaged on passage through the gut. Closely related species, such as the salmon (*Salmo salar*) and trout (*Salmo trutta*) often cannot be separated because their vertebrae and scales are very similar. Frequently it is only possible to identify fish to family level. However, some information can be gained by comparison with reference collections (made up from the scales, vertebrae and jaw apparatus of known species) and a knowledge of the potential prey species available obtained by electro-fishing or rock-pool sampling. Methods are also available for relating the length of vertebra to length of the fish from which it originated, thus allowing the size to be estimated.

Fish prey taken in North America A great variety of fish species have been recorded as having been consumed by mink in North America. The family Cyprinidae shows the greatest diversity of fish inhabiting fresh water and, as might be expected, more members of this group are taken than any other. Included among the cyprinids found in a study of mink from New York State (Hamilton 1959) are the horned dace (*Semotilus atromculatus*), blacknose dace (*Rhinichthys atratulus*), common shiner (*Notropis cornutus*) and golden shiner (*Notemigonus crysoleucas*). Both brook and brown trout were represented. Mink from Long Island had fed principally on mummichog (*Fundulus heteroclitus*), common pumpkinseed (*Lepomis gibbosus*) and smallmouth bass (*Micropterus dolomieui*). The suckers found were all individuals of *Catostomus*. Members of the perch family were also occasionally taken.

Fish prey taken in Europe A wide range of species are taken, although this may simply reflect the more adequate diagnostic keys to fish remains that are available there compared to North America.

Probably the most thorough analysis of fish predation by mink in freshwater habitats was that undertaken by Wise *et al.* (1981) in Devon. Two habitats were chosen, a eutrophic lake (Slapton Ley) and an oligotrophic moorland river (Dart/Webburn). At the lake, fish made up 31% of the mink's diet. A cyprinid, probably the roach (*Rutilus rutilus*) was the main fish prey taken. Eel (*Anguilla anguilla*) was the second most important fish prey followed by perch (*Perca fluviatilis*). Pike (*Esox lucius*) were taken infrequently. On the rivers, fish were of lesser importance to mink (24.8% of the diet). Salmonids (brown trout, sea trout and salmon were all present in the river) made up 22% of the diet, and were relatively more important in autumn and winter. Eels were also present, but rarely taken, as were bullheads (*Cottus gobio*). On a chalk stream, Chanin & Linn (1980) found eels and cyprinids to be important prey.

It appears that lake habitats are less optimal for the mink's hunting strategy since it is forced to search open water. Circumstances where laboratory studies have suggested it should be less efficient at hunting. Quite possibly the mink confined their hunting behaviour to the reedy margins of the lake.

In Scotland a combined analysis of mink stomach contents and scats by Akande (1972) indicated salmonids (brown trout and salmon) to be the main fish prey consumed. Some grayling (*Thymallus thymallus*) were also taken. Eels, although plentiful in the rivers, were rarely predated in this study. An extensive study by Jenkins & Harper (1980) showed that eels could be a major prey item in the diet of mink from Deeside. Similarly, Cuthbert (1979) demonstrated salmonids and eel to be the major dietary constituent of mink from the Rivers Tweed, Urr and Sheeoch in Scotland.

In Sweden (Gerell 1967b), the fish taken commonly by mink are *Salmo* and bullhead. Perch, pike, burbot (*Lota lota*), rudd (*Scardinius erythrophthalmus*), ide (*Leuciscus idus*), bream (*Abramis brama*) and minnow (*Phoxinus phoxinus*), also occurred occasionally in the diet. On the River Skog in southern Iceland, Skirnisson (1990) found salmonids and sticklebacks (*Gasterosteus aculeatus*) to be the principal prey.

On the coast, fish can assume a very significant proportion of the diet. On the Galloway coast of Scotland, 67% of scats collected contained fish vertebrae (Dunstone & Birks 1987). Six species predominated in the diet. The blenny (*Lipophrys pholis*), rockling (*Ciliata mustela*) and sea-scorpion (*Taurulus bubalis*) were the most commonly taken fish (Table 6.5). The sea-scorpion and the butterfish (*Pholis gunnellus*) were taken more frequently by

TABLE 6.5 *The occurrence of fish species in the diet of mink and in rock pools where they fish.*

Fish species	Percentage	
	Scats	Survey
Blenny	37.2	17.9
Five-bearded rockling	21.6	10.3
Long-spined sea-scorpion	19.1	43.5
Common eel	7.4	5.1
Butterfish	5.9	15.4
Goby	6.0	7.0

Source: Dunstone & Birks (1987).

mink than was to be expected from our sampling of rock pools, perhaps indicating that these species may be more vulnerable to predation by mink. In Iceland, Skirnisson (1980) also found fishes from intertidal and shallow waters to be important. Another species of sea-scorpion, *Myxocephalus scorpio*, made up 42% of the diet, but butterfish, rockling and lumpfish (*Cyclopterus lumpus*) were considerably less important than in Scotland. Most of the coastal fish taken were rock pool inhabiting species that rely on hiding away and camouflage to evade predators rather than speed of escape or schooling behaviour.

Size of fish taken

Occasionally it has been possible to estimate the size of the fish taken. Cuthbert (1979) considers that most of the Scottish salmonids taken by mink were small: 72% were less than 12 cm, 25% between 12 and 25 cm and only 3% were greater than 25 cm. Chanin (1981) has found similar size ranges to be taken by mink on the River Teign in Devon. The size ranges of eels can be highly variable, from 2 to 25 cm and weighing from 2 to 200 g (Akande 1972). The maximum size of salmonids and eels taken by mink in Devon (Wise et al. 1981) were 40 cm and 50 cm, respectively. Perch ranging from 6 to 12 cm and pike up to 60 cm (although more frequently of between 30 and 50 cm) were taken.

AMPHIBIA

Amphibia, especially frogs may be seasonally or locally important, but generally only make up a minor proportion of the diet. Only in one study (Korschgen 1958) were frogs the major prey group, and even then they made up only 25% of the diet. These were mainly the bullfrog (*Rana catesbeiana*), leopard frogs (*Rana pipiens*) and the spring peeper (*Hyla crucifer*) (Hamilton 1959). In North America, frogs were possibly taken as torpid individuals from pond bottoms, where they overwinter. Other more aquatic species such as salamanders (*Desmognathus* sp., *Eurycea* sp. and *Ambystoma maculatum*) were much less important in the diet (Hamilton 1959). Eberhardt (1977) records leopard frogs (*Rana pipiens*) and the larvae (tadpoles) of the tiger salamander (*Ambystoma tigrinum*) in the diet of marsh inhabiting mink.

In the UK, frogs (*Rana temporaria*) have only rarely been shown to be important. During one year of their three year study, Wise *et al.* (1981) found frogs made up 9.6% of the mink's diet. *Rana temporaria* and *R. arvalis* have also been shown to be of importance in some Swedish studies (Gerell 1967b, Erlinge 1972). The peak in amphibian predation was correlated with the occurrence of spawning, but overall Amphibia are considered to be a second-class food only taken when other prey are not available.

REPTILES

Smooth green snake (*Opheodrys vernalis*) and garter snake (*Thamnophis* spp.) remains have been found at mink dens (Eberhardt 1977). Garter snakes, red-bellied snake (*Storeria* sp.) and a painted turtle (*Chrysemys* sp.) were also recorded by Hamilton (1959). In England, Day & Linn (1972) recorded the remains of lizard limbs in mink stomachs.

CRUSTACEA

In America, the crustacean component of mink diet is mainly represented by the crayfish (*Cambarus* sp.). The extent to which they may be taken can vary from 6% (Sealander 1943) to 19% (Korschgen 1958). The latter author suggests that Crustacea can become particularly important during drought conditions.

In Britain, crayfish (*Astacus astacus*) are only locally abundant, and hence have not figured prominently in many published descriptions of the mink diet. Nevertheless, on the Rivers North Tyne and Wansbeck in Northumberland, I have frequently found scats coloured red by the remains of this prey. In Sweden, crayfish can be an important item, especially in the warm months (Gerell 1967b), whereas their reduced activity during winter makes them less available. The seasonal nature of their predation will be discussed later.

For coastal mink in the UK, the shore crab (*Carcinus maenas*) may be seasonally important (Dunstone & Birks 1987). These are probably caught on the shore or in shallow rock pools and do not require underwater pursuit by mink. On Vancouver Island, British Colombia, Hatler (1976) found crabs to be the most intensively used prey group. Two species were of particular importance, the red crab (*Cancer productus*) and the northern kelp crab (*Pugettia producta*), although other species were also taken. Hatler (1976) assessed the importance of the crabs from direct observation of foraging behaviour and analysis of the contents of mink middens. The red crab predominated in all seasons and the kelp crab became an important constituent of the winter and spring diet. There appeared to be a preponderance of female kelp crabs in the midden remains; perhaps the mink experienced difficulty in handling the larger, more pugnacious males. Mink appear to select crabs of around 6 cm carapace width. Most of the smaller specimens remain hidden in sea-weed, and larger specimens are rarer and presumably more difficult to handle.

OTHER INVERTEBRATES

Hamilton (1959) has shown that water beetles (*Dytiscus*) and their larvae, tipulid larvae, *Neuroptera* larvae and large *Plecoptera* nymphs can be commonly taken. However, apart from *Dytiscus*, these invertebrates are a relatively unimportant component of mink diet and are not considered further.

In coastal habitats in Scotland and Iceland, large numbers of the sea slater (*Ligia oceanica*)—a small terrestial crustacean resembling a large woodlouse—were taken even though these must be of dubious food value (Skirnisson 1980, Dunstone & Birks 1987). Hatler also found sea slaters (*Ligia pallasii*) in the diet of Vancouver Island mink which he thought were taken at night when the sea slaters emerged from their crevices and moved into the intertidal zone to feed.

At extremely low tides, the mollusc abalone (*Haliotis kamschatkana*) become available to Canadian mink. Shells up to 11 cm diameter were found in mink food caches by Hatler (1976), who also reports that sea urchins may also be taken.

Many carnivores (e.g. European badgers and foxes) are known to forage extensively on earthworms. Few authors have looked specifically for this prey in mink scats, but their presence has been demonstrated (Chanin & Linn 1981). Skirnisson (1980) reported bumblebee (*Bombus jounelis*) remains and also found bee nests that had been dug up by mink.

CARRION

Only in one UK study has this been considered to be a major source of food for mink (Dunstone & Birks 1987). A large variety of corpses were found at a coastal site in southern Scotland. Some of these were known to be taken by mink, for example, when shotgun pellets and rabbit fur turn up in the same scat! A profusion of sea-bird carrion was available in the vicinity of their breeding colonies and washed up on the shore after storms.

FACTORS AFFECTING DIETARY COMPOSITION

VARIATION WITH LOCATION AND SEASON

As might be expected from any opportunistic predator, the diet will reflect what is commonly available to the mink in a particular locality. Figure 6.10 demonstrates how, even within a small study area the diet can vary considerably depending on the relative availability of different prey groups.

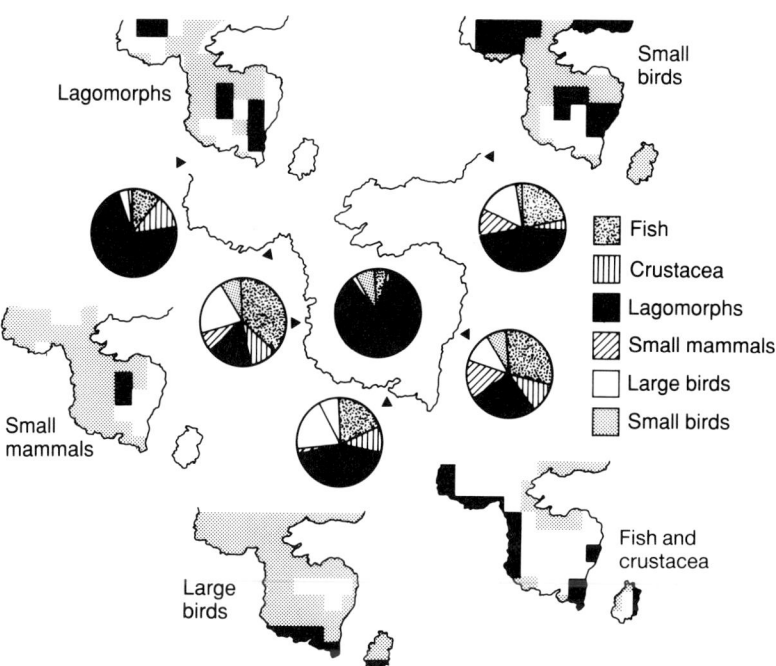

FIGURE 6.10 *Local variations in the diet of coastal mink in Scotland. Pie charts represent dietary analyses based on scats collected in the habitats indicated. The intensity of shading on maps is related to the relative abundance of prey types in the study area.*

VARIATION WITH SEASON

The occurrence of different prey types in the mink's diet varies with the seasonal changes in prey abundance and behaviour, which are related to temperature and breeding activities. In the winter months, fish are frequently found to be the major food source. Gerell (1968) suggested that this was due to their greater susceptibility to capture in colder water. This is clearly demonstrated in Fig. 6.11 where the change in the percentage occurrence of aquatic prey in the diet of Swedish mink on the river Rönnea can be seen to

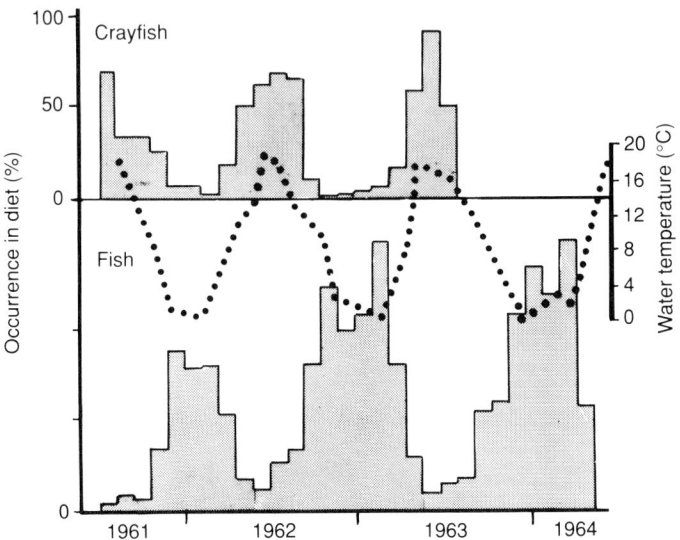

FIGURE 6.11 *Percentage occurrence of crayfish and fish in the diet of mink in relation to water temperature. Source: Gerell (1967b).*

fluctuate with the water temperature. While this may be the case for those species that rely on speed to evade a predator, it is not an adequate explanation for why the slower swimming, rock pool inhabiting species also features so commonly in the diet of coast living mink during winter. Dunstone & Birks (1987) suggest that this may result from the decreased availability of alternative prey, principally rabbits. Chanin & Linn (1981) noted that predation on eels was twice as great during summer and winter compared to spring and autumn, but failed to find an adequate explanation since this prey is thought to be relatively inactive from November to March (Perrett 1958). Wise *et al.* (1981) suggested that eels become torpid in winter and hide in the mud of lakes where they are protected from mink predation.

In Sweden, crayfish became increasingly important in summer (Fig. 6.10) as their mobility, and hence vulnerability, increases with rising water temperature. Compared with fish, the crayfishes' more secretive way of life serves to reduce their vulnerability in winter. In Scotland, Dunstone & Birks (1987) have also demonstrated the seasonal nature of shore crab predation (Fig. 6.12), particularly during the late summer when they migrate on shore for pairing (Naylor 1962).

Rabbits were the single most important prey of coast living mink in Scotland (see Fig. 6.12), and the seasonal cycle of predation matched closely the abundance of rabbits and hares on the study site as determined from counts (Fig. 6.13). In North America muskrats were taken predominantly during the drought season, so-called low-water winters, when their populations become overcrowded in the few remaining bodies of water

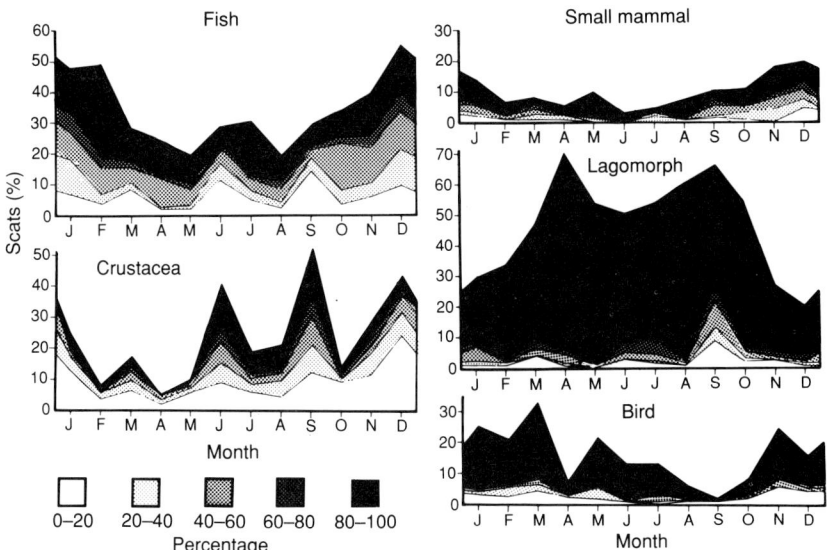

FIGURE 6.12 *Seasonal variation in the diet of coast-living mink in Scotland. Source: Dunstone & Birks (1987).*

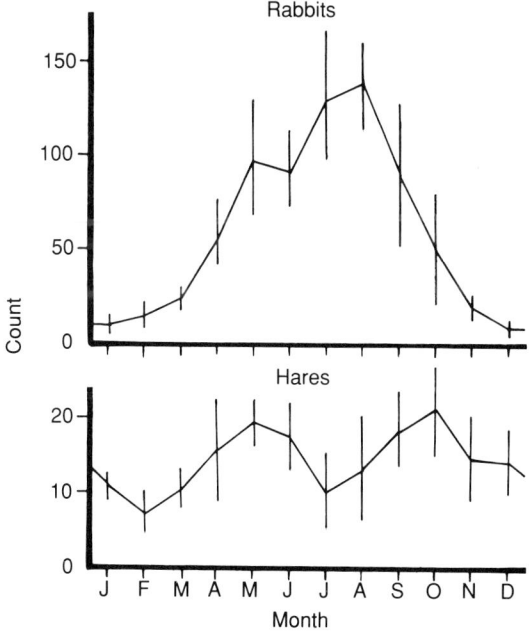

FIGURE 6.13 *The abundance of rabbits and hares as determined from visual counts at dusk. Source: Dunstone & Birks (1987).*

available. Surplus individuals then become more accessible to mink. On the prairie marshes of North Dakota, where fish and Crustacea were rare, Arnold & Fritzell (1987) found mammals (muskrats and ground squirrels) to be predominant in the diet during May and April, but were gradually replaced by coots and later (in July) by ducks and ducklings. The increase in avian predation resulted from their decreased mobility arising from the requirement of incubation, brood rearing and the moult. Similar seasonal fluctuations in the use of waterfowl have been noted in Europe. Wise et al. (1981) noted that the importance of ducks, coots and moorhens as prey increased in the winter and summer and decreased in the spring and autumn (Table 6.6). It

TABLE 6.6 *Predation on ducks (Anseriformes) and coots and moorhens (Ralliformes) on Slapton Ley, Devon.*

Season	Visual estimate (%)		% in Mink diet	
	Anseriformes	Ralliformes	Anseriformes	Ralliformes
1975				
Winter	86.3	13.7	16.8	83.2
Spring	88.6	11.4	34.2	65.8
Summer	70.4	29.6	41.9	58.1
Autumn	75.8	24.2	82.6	17.4
1976				
Winter	85.2	14.8	27.1	72.9
Spring	81.0	19.0	37.6	62.4
Summer	65.2	34.8	19.8	80.2
Autumn	66.0	34.0	59.7	40.3
Overall mean	76.4	23.6	36.4	63.6

Source: Wise et al. (1981).

is evident that mink are selecting coots and moorhens at this time. Gerell (1967b) also records seasonal predation on these avian groups. A higher proportion of coots was taken during the breeding season, whereas predation on ducks was highest in the autumn. On the coast, little seasonal variation was evident in the predation of sea-birds, possibly because the data include many different species of widely different seasonal availability and the mink's use of carrion.

These field investigations clearly demonstrate that the food habits of mink vary with the season and habitat due to differences in the abundance and vulnerability of the potential prey available. In general, fish are the predominant prey over winter and spring. Mammals are the second most important group, while crustaceans and birds are not commonly taken at this time. During summer the diet is diverse as breeding birds and an abundance of mammals are available. Crustacea too are vulnerable at this time. By autumn, bird availability has usually decreased, but the summer population explosion of mammals can still be exploited, with some species continuing to breed into autumn. As water temperatures drop, the availability of fish rises again, to dominate the diet over winter. Hence, the greatest transitions in diet occur from spring to summer and from autumn to winter.

In the north-western USSR (Karelia, Leningrad and Pskov regions) the percentage of scats and stomachs of European mink containing mammalian prey (predominantly muskrats, voles, shrews and moles) fell from 33% in winter to 18% when there was no snow cover. In contrast, the proportion of birds in the diet rose from 4.8% to 20.3%, frogs rose from 25.4% to 72.9% and fish from 6.7% to 26.4% during winter to summer (Danilov & Tumanov 1976).

SEX DIFFERENCES IN FEEDING BEHAVIOUR

Analysis of just over 100 mink guts from winter-trapped mink in Michigan led Sealander (1943) to comment that '. . . there was an apparent discrimination by sex as to the size of prey items taken'. More muskrats were taken by male mink then by females, and there was also a sex-related difference in the numbers of small mammals and fish found, with females apparently taking smaller prey than males. Sealander (1943) attributes this to the size difference in male and female mink, suggesting that 'the larger males may prefer to satisfy their appetites more quickly'.

The sex difference only applies to muskrats and not to cottontail rabbits, the former perhaps being more formidable prey for a female mink. The wider ranging movements of the male mink may also lead to him encountering more muskrats than does the female.

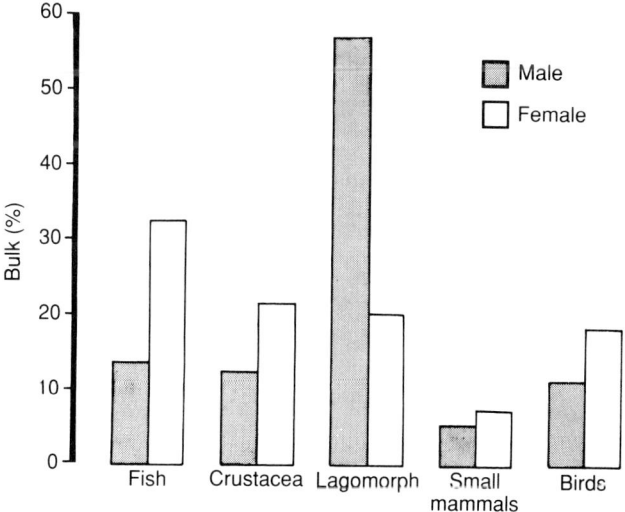

FIGURE 6.14 *The diet of male and female coast-living mink in Scotland. Source: Birks & Dunstone (1985).*

Although a number of dietary investigations conducted in the UK have relied on gut contents from trapped animals, where the sex of the mink should have been known, only one attempt has been made to quantify sex differences in diet. Birks & Dunstone (1985) set out to examine the effect of sexual dimorphism of body size on the diet of mink by analysing 1024 scats collected from radio-tagged individuals. The hypothesis under investigation was whether male mink use prey which are relatively inaccessible to females or vice versa. In the coast living population studied, males averaged 1.74 times the female body-weight. The relative contributions of the five prey groups to the diet of each sex are illustrated in Fig. 6.14. Fish and crustacean prey were taken more frequently by females. Bird prey consumption also differed, but the results were biased by inclusion of data from a female known to be taking sea-bird carrion. There was no sex-related difference in predation of small mammals. But the most striking difference between male and female mink was observed in their predation of lagomorphs. Males were observed to attack adults. Even when the female took such 'large' species these tended to be juvenile individuals. Ireland (1990) observed a female digging out a rabbit breeding stop. Of nine rabbit carcasses found in a female mink's den during the summer of 1983, only one was adult. This should not be taken to imply that female mink cannot kill rabbits, after all, weasels and stoats which are both considerably smaller than female mink do so occasionally. A more likely explanation is that because of the high energy cost of killing a rabbit, more profitable alternative prey are taken.

The size effect is more clearly demonstrated when the data are categorized into 'large' and 'small' items (Fig. 6.15), female mink emerge as predators of

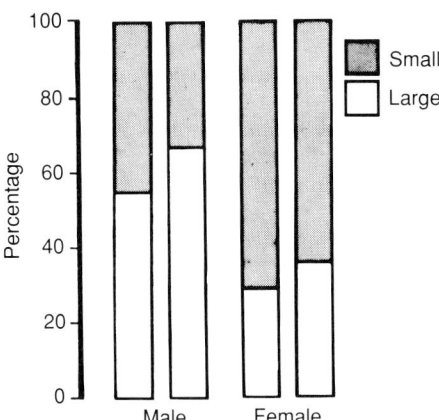

FIGURE 6.15 *The proportion of large and small prey in the diet of male and female coast living mink. Source: Birks & Dunstone (1985).*

prey items which are predominantly smaller and generally 'aquatic' (Fig. 6.16). There were seasonal differences in the consumption of 'small' versus 'large', and 'terrestrial' versus 'aquatic' prey. A potential problem arises in this interpretation since young male mink overlap in size with females during the summer and autumn. Ireland (1990) has recently shown that it is indeed a

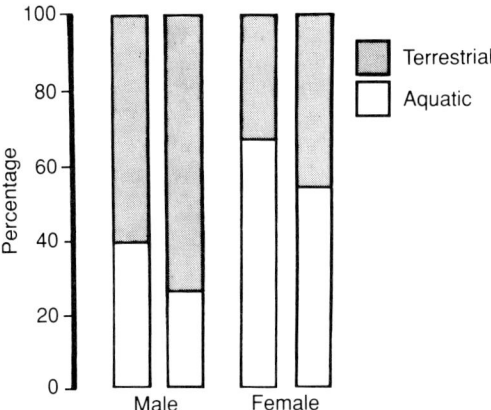

FIGURE 6.16 *The proportion of terrestrial and aquatic prey in the diet of male and female coast living mink. Source: Birks & Dunstone (1985).*

sex difference in diet rather than body size *per se*, since small males also prefer larger prey species. In general, larger prey are more profitable on energy grounds, as long as handling costs are not too high. A fully grown rabbit can sustain an adult male mink for 2–3 days (Linn & Birks 1981, Dunstone & Ireland 1989). The rabbits burrowing habits make them vulnerable to predation underground where speed of pursuit is not vital. The different head shape of male and female mink and its consequences for larger canines and carnassial teeth and larger more powerful jaw muscles may improve the males' chance of holding on to larger prey (Gregory 1987).

While the relatively large males may consume foods unavailable to the females, the reverse is much less true. Although rabbit prey were heavily exploited by males in some seasons, they also take large quantities of seashore prey during the autumn, even though lagomorphs are still abundant at that time. A possible explanation lies in the seasonal increase in availability of crabs. Females did not harvest the crabs to the same extent as males. We could not determine whether the differences in diet were related to differences in availability, or preference, or competitive interaction. Birks (1981) has clearly demonstrated that female mink are subordinate to males outside of the breeding season, possibly restricting the females access to this resource during the period of crab abundance.

Hence, although dietary differences are apparent between the two sexes, they are not so extreme or fixed as to be the main driving force in the

INDIVIDUAL DIFFERENCES IN DIET

At the individual level, some mink in some populations may specialize in feeding on particular prey types for several months at a time, and they could then have a considerable local impact on these species (Dunstone & Ireland 1989). Figure 6.17 shows dietary results from individual coast-living mink.

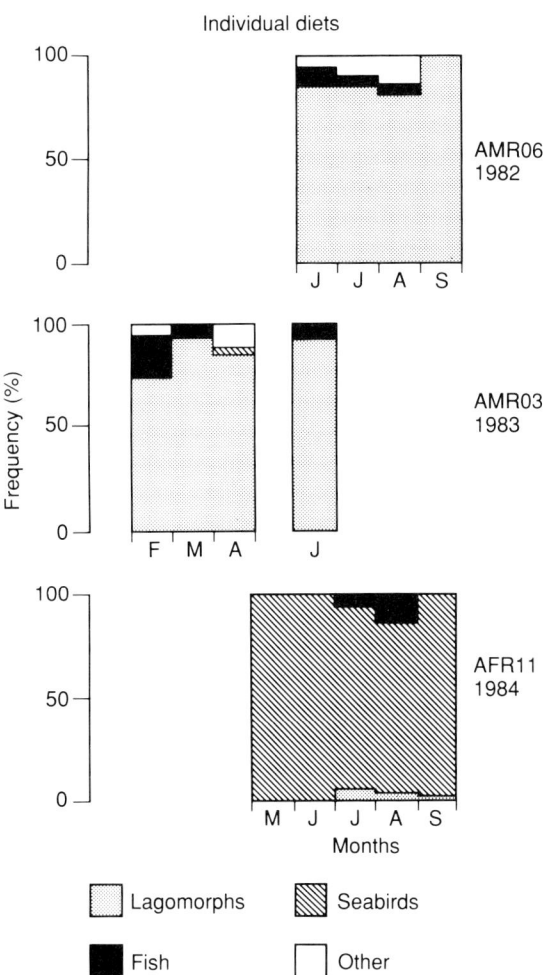

FIGURE 6.17 *The diets of individual coast living mink.* Source: Dunstone & Ireland (1989).

The data are from two males and one female resident in the area for a considerable time and whose movements and activity were monitored by trapping and radio-tracking. During this time faecal material was regularly collected from the dens they occupied. The two males appeared to direct their predation mainly towards rabbits. For one male this coincided with the peak availability of rabbit prey in this area, but for the other, rabbits were also taken when they were considerably less abundant. The female raised three litters in consecutive years on a small off-shore island where her food requirements were met by predation upon a colony of herring gulls (*Larus argentatus*) which nested there, and from scavenging carcasses.

ECOLOGICAL IMPACT OF MINK ON PREY POPULATIONS

All considerations of the diet of mink should be considered for their ecological impact on the numbers of the prey animals they are taking, and whether or not these are 'sensitive species'.

The biological value of a single prey item is a function not only of its nutritional quality, but also of its abundance relative to other prey types, and the energy expended in search, pursuit, capture and handling. The relative digestibility of different prey types and the proportion of indigestible remains, for example bone, fur and feathers, will also vary, but these may need to be taken to provide roughage in the diet.

Information is available on the daily energy requirements of the mink, and data from laboratory feeding trials allow the value of the different prey types to be assessed. From this information it is possible, albeit approximately, to estimate the mink's daily prey requirement.

The estimated weight specific energy requirements of mink range from 635 kJ kg$^{-0.75}$ day^{-1} (Cowan *et al.* 1957) to 1299 kJ kg$^{-0.75}$ day^{-1} (Gregory 1987). Given a mean weight for wild mink of 1200 g and 700 g for males and females, respectively (Birks & Dunstone 1985), then this would suggest a daily energy requirement of approximately 1500 kJ for a male and 1000 kJ for female.

Hence, it is possible to estimate the weight of prey required to fulfil the daily energy requirement of a mink, and therefore the quantity of each prey type that would be required to sustain the predator. Table 6.7 shows the approximate daily consumption of various prey species by male and female mink. No attempt is made to take account of varying energy demands incurred by the animal at different times of the year resulting from varying activity levels (e.g. rearing a litter) or the effect of ambient temperature. These values should be considered as absolute since predators often do not get the total amount of energy potentially available from their prey. However, they do give an indication of the relative value of different prey types.

Undoubtedly the large prey items such as rabbits and large birds are more economic prey since they could sustain the mink for a number of days. Small

TABLE 6.7 *Energy content of prey types commonly taken by mink and the theoretical duration for which individual items would sustain the predator.*

Prey type	Mean wet weight (g)	Calorific value (kJ(g dry wt)$^{-1}$)	Energy content (kJ)	Longevity of carcass (days)	
				Male	Female
Mammals					
Rabbit (*Oryctolagus cuniculus*)	1500[e]	20.69[b]	9311	6.3	9.5
Brown hare (*Lepus capensis*)	3540[e]	23.75[b]	25 222	17.0	25.7
Field vole (*Microtus agrestis*)	35[e]	21.0[a]	222	0.15	0.23
Birds					
Starling (*Sturnus vulgaris*)	75[f]	24.33[a]	547	0.37	0.56
Domestic fowl (*Gallus domesticus*)	1341[c]	23.64[c]	9828	6.62	10.02
Fish					
Eel (30cm) (*Anguilla anguilla*)	290	12.93[c]*	1126	0.76	1.15
Trout (20cm) (*Salmo trutta*)	113[f]	32.36[c,g]	1097	0.74	1.12

Source: [a]Moors (1977), [b]Myrcha (1968), [c]Gregory (1987), [d]Sidwell (1974), [e]Corbet & Southern (1977), [f]Frost & Brown (1967), [g]Murray & Burt (undated pamphlet.) *Wet weight.

mammals (of vole size) or passerine birds would have to be obtained in considerable numbers to meet the energy needs of the predator; mink could have a significant impact on their populations if they were specialist predators on these prey. That they occur rather infrequently in the diet suggests that they are uneconomic items.

In terms of energy content, fish prey provide a better yield than do terrestrial animals. However, their net values will be less since recent studies (Stephenson *el al.* 1988) have demonstrated that the capture of aquatic prey will require a considerably greater level of energy expenditure to provide an equivalent level of calorie intake.

Arnold & Fritzell (1987) carried out similar calculations to determine the impact on waterfowl populations by prairie mink in Manitoba. They estimated that the daily requirement of male mink in their study area was 0.18 kg of prey per day, or 22 kg over the waterfowl breeding season. Each male mink could have achieved this intake by consuming, on average, 2.7 kg of adult ducks (3 mallard), 1.5 kg of ducklings (15-week-old mallard) and 0.9 kg of eggs (18 eggs) during each breeding season. This was rated to be equivalent to only 1% of the waterfowl population and its potential reproductive output. In North Dakota, Eberhardt & Sargeant (1977) found female mink to have a considerable impact on the waterfowl population because they were raising young at the time and had a higher absolute prey requirement than male mink did. Data acquired during feeding trials with mink housed on farms showed that only 6% more food is required during gestation, but that the females feeding requirements may increase to twice her normal intake before weaning the kits. In the wild, Ireland (1990) suggests this might rise to five times the prey requirement where a wild population is concerned.

The major concern about predation by feral mink relates to their effect on native bird species. In this context it must be remembered that mink are territorial and the size of the territory is influenced by two factors, the habitat quality and the presence or absence of other mink. Because of this spacing mechanism, any local abundance of food or seasonal glut will only be available to the territory holders, thus allowing only a small number of animals to share the food, while preventing mink from surrounding areas gaining access. Thus, for example, on Ross Island where there were abundant ground-nesting herring gulls available for most of their breeding season only one female mink preyed on the colony because it lay within her territory. This sea-bird colony has continued to expand despite the annual predation pressure from these animals (Birks 1986).

CHAPTER 7

Lifestyle

HABITAT REQUIREMENTS

The mink is a semi-aquatic mammal that is usually found associated with stream and river banks, lake shores, fresh and saltwater marshes and marine shore habitats. Although the range of the mink invariably includes a wide variety of water systems, not all are equally suited to its style of semi-aquatic activity.

We have already noted that the mink's adaptations suit it for hunting in shallow, slow-moving or stationary bodies of water. Burgess (1978) observed that the likelihood of finding evidence of mink decreased markedly as stream flow increased. Furthermore, as Korschgen (1958) has pointed out, aquatic foods become more important in the diet in areas where the water depth is shallow. In a coastal habitat, sheltered rocky shores are ideal for mink.

It is important to distinguish between the suitability of different aquatic habitats, since the biological richness will also affect their suitability as a foraging site.

Eutrophic waters are characterized by being shallow with a high organic content and high concentrations of nutrients for plant growth. These may be either natural or caused by agricultural run-off. The shoreline vegetation is usually prolific. In contrast, oligotrophic waters are typically deep with scarce shoreline vegetation, and the waters have a low plankton and organic content. Eutrophic lakes and rivers have a greater potential for productivity than do oligotrophic waters, and may therefore be capable of supporting a larger population of mink.

The distribution of mink can be affected by the amount of cover available (Allen 1983, Mason & Macdonald 1983), particularly mature willow (*Salix* spp.), saplings and shrubs. The habit of pollarding willows can provide good den sites. Even un-pollarded willows provide refuges for mink, since their trailing branches often accumulate debris where they touch the water.

Most mink activity in North America and Europe occurs within 100–200 m from water. For example, in a study based in Michigan, the majority occurred within 30 m of the water's edge (Burgess 1978). Melquist et al. (1981) also recorded most mink activity within 200 m of river margins. The farthest that coast-living mink in Scotland regularly moved from the mean high water mark was 500 m (Ireland 1990).

Apparently the European mink occurs primarily along small forest streams, especially those that are less likely to freeze completely during winter. In contrast to the American mink, this species is said to occur infrequently on large rivers, and then preferentially at the mouths of small tributaries. They are rarely found along lake shores (Danilov & Tumanov 1976).

Mink dens are preferentially located near the water's edge, depending on the availability of den sites. Birks & Linn (1982) recorded all dens on the River Exe and at Slapton Ley (a lake) to be within 10 m of the water, and most were within 2 m. On the River Teign, dens were about 130 m from the water's edge; however, all of these were in rabbit burrows. The dens of coast-living mink were usually within 70 m of the high tide line and some temporary ones were below it (Ireland 1990)!

Russian work has shown that the European mink use nests and shelters located under tree root systems and rock piles. The homes of muskrats (which are now feral in several parts of Europe) were also extensively used. Unlike the American mink, *M. lutreola* seems to favour lining its den with dry grass, leaves, feathers or rodent fur. Danilov & Tumanov (1976) record the internal diameter of the den to be from 20 to 35 cm.

VEGETATIVE COVER

Mink, throughout their native and feral range, are most commonly associated with wooded, brush or scrub cover adjacent to aquatic habitats, especially during those seasons when they are most dependent on aquatic foods. Mink generally avoid open or exposed areas. They are not swift in pursuit, and are better suited to hunting prey in confined spaces. Ireland (1990) noted that the speed of a mink he observed pursuing a rabbit unsuccessfully across pasture was equivalent to the speed at which it returned to the security of cover after the hunt.

In America, marshes containing dense stands of saw-grass (*Cladium jamaicense*) support high densities of mink. The cypress-tupelo swamps characteristic of southern Louisiana provide an abundance of food resources and den sites. There is evidence that semi-permanent wetlands provide better opportunities for mink. In Sweden, Gerell (1970) characterized optimal mink habitats as small oligotrophic lakes surrounded by stony shores and marsh vegetation. Wetlands with irregular shorelines provide more favourable habitats than those that are straight and exposed. Similarly, habitats associated with small streams are preferred to large broad rivers.

Mink are most common along streams where there is an abundance of debris to provide cover and pools of still water holding fish. Melquist et al. (1981) found log jams to provide excellent foraging opportunities for mink since they provided shelter for aquatic prey, aerial vantage points overlooking the fish pools, and security for the mink.

It is possible to enhance the habitat for mink. Burgess & Bider (1980) improved a section of stream by building small dams causing the formation of pools and riffles. Bankside cover was then provided close to these pools, and large rocks submerged to provide refuges for the fish prey. This resulted in a doubling of the trout population and of the biomass of crayfish in the modified section compared to a control area which was not altered. They recorded a 52.5% increase in mink activity in an area surrounding the section that had been improved. This was due to the increase in crayfish population rather than trout, since the latter was not an important prey species in this study. These results are in accordance with laboratory observations made by Dunstone & O'Connor (1979a,b) who found that mink forage inefficiently in open water.

To a limited extent, mink can tolerate human disturbance. Birks (1981) radio-tracked a mink along sections of the River Exe as it flowed through the town of Exeter, Devon. Many people have recorded the presence of mink in suburban areas. Racey & Euler (1983) investigated the effects of lakeside cottage development in Ontario on the activity of mink. The clearance and modification of vegetation during lot-clearing and road building disturbs mink, either directly or indirectly, by affecting prey populations. Tree and shrub vegetation were removed as were underwater snags, aquatic and overhanging vegetation. As might be expected, mink activity—based on the number of scats found—was greatest on areas with little or no development. But, although opportunities for shoreline foraging were reduced, fish predation seemed to increase in developed areas. Small fish were thought to be attracted to docks built by cottagers for mooring their boats, and this led to an increase in their local abundance. Simplification of the shoreline by removal of aquatic snags and vegetation may have led to the creation of a patchy environment with local concentrations of fish prey.

Mink are opportunistic foragers and switch between habitat types in response to changes in the availability of their prey. An extensive analysis of habitat use was carried out by Mark Ireland, Johnny Birks and myself (in prep.) on a coast-living population of mink in southern Scotland (see Figs 7.1 and 7.2). Here, the manner in which mink partitioned their activity between habitats reflected the availability of dens and the use of particular areas for foraging.

The heterogeneous nature of the coastal habitat—steep cliffs, rock pools, boulder-strewn beaches fringed by forestry plantations, agricultural pasture and scrub—provided a wide diversity of prey types. Female mink made greater use of the shore compared to males, which almost certainly reflected their greater dependence on aquatic prey. In contrast, males spent more time foraging in the fringing rocks and scrub, where they fed particularly on rabbits. Although rabbits were also available on pasture, the open nature of this habitat meant that mink could not hunt them there, hence reducing its value as a foraging area. When mink travelled along the coast they tended to

FIGURE 7.1 *The Ross Peninsula and Ross Island, southern Scotland. (Crown copyright)*

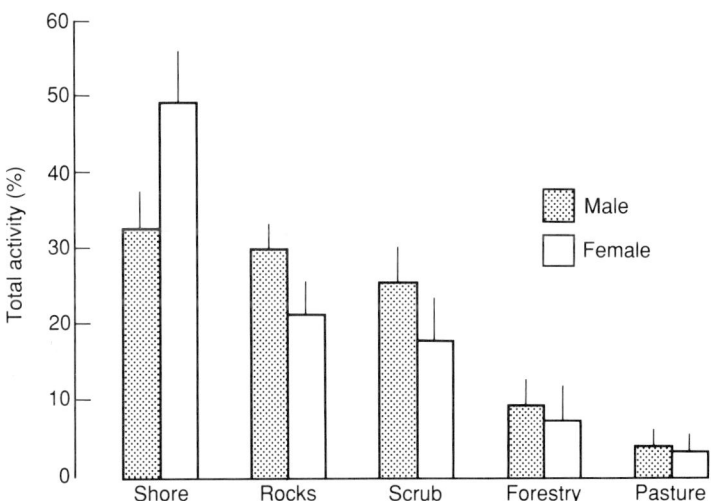

FIGURE 7.2 *Habitat use by coast-living mink. The proportion of out-of-den activity spent in each of the five main habitat types by male and female mink. Source: Ireland (1990).*

confine their activity to the rocky zone just above the high tide level, probably because these areas provided abundant shelter and security. Because it was the focus of considerable activity, this zone was important for the mink's social behaviour since it was here that encounters with other mink were likely to occur.

DEN SELECTION

Mink do not use the whole of their range evenly (Gerell 1970); certain core areas are used more frequently than others. Females particularly show the greatest intensity of use in areas with occupied dens. Within the home range mink will use several dens for shelter, concealment and rearing young. Such dens may be used for less than 1 day, or up to 40 days by a female with kits. Mink do not excavate their own dens, although they have been known to enlarge those of other animals. During his research studies, Birks (1981) excavated the den of one mink and found a nest chamber lined with woodrush and separate areas for use as a larder and latrine (Fig. 7.3). The

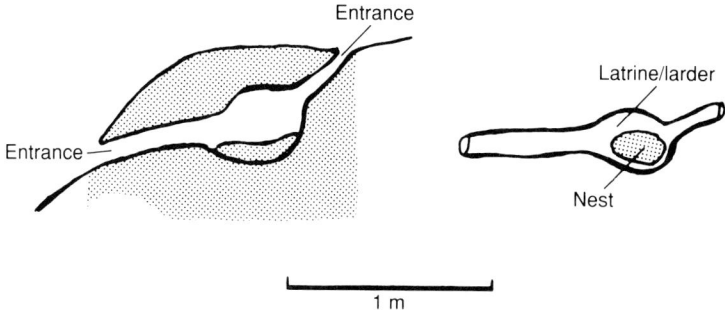

FIGURE 7.3 *Diagrammatic representation of an excavated mink breeding den. Source: Birks (1981).*

most commonly used den sites in riparian habitats are cavities between tree roots at the water's edge, or in rock piles. Table 7.1 categorizes the minks' choice of dens in three habitat types.

The minks' dependence on burrows and cavities, and their restriction of foraging behaviour to the vicinity of dens, suggests that an adequate supply of suitable sites is crucial (Birks & Linn 1982). In areas where there is a surfeit of potential sites (e.g. exposed tree root systems), mink may select particular

TABLE 7.1 *The types of dens (%) used by mink in three types of habitat.*

Den type	Coastal	Percentage Lacustrine	Riverine
In or beneath waterside trees	0	29	42
Rabbit burrows	35	3	18
Other holes and crevices	60	52	26
Human artefacts	5	10	7
Above ground (scrub, brush)	0	6	7
No. of dens	85	31	57

Source: Dunstone & Birks (1987).

dens because of their vicinity to preferred foraging sites. In contrast, in areas where prey is more uniformly abundant (e.g. marshy areas), the distribution of den sites may itself determine where the mink forages.

In Idaho, Melquist *et al.* (1981) demonstrated that mink favoured log-jam dens, but rock crevices with small openings were also used. Abandoned or seldom-used muskrat burrows can also be important if available (Eberhardt & Sargeant 1977). In Sweden, Gerell (1970) recorded mink using between two and five dens, the mink changing dens on successive nights, generally choosing the next nearest den available which was, on average, 500 m distant (see Fig. 7.4). Similarly, Ireland (1990) found that 60% of mink changed dens overnight. The greatest frequency of change was by males during the rutting season, while females with kits only rarely changed dens. In North Dakota,

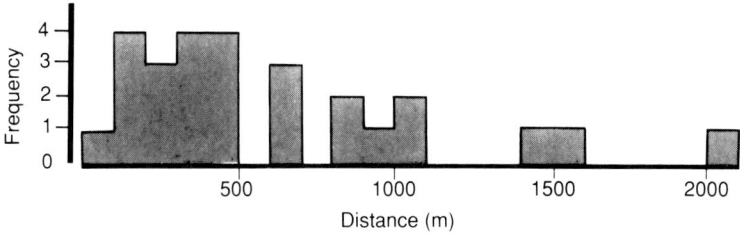

FIGURE 7.4 *Frequency distribution of distances between dens. Source: Gerell (1970).*

Eberhardt & Sargeant (1977) found that breeding dens were used only during May to July, with the maximum duration of occupation being 40 days when the kits were young. Subsequently, dens were used for 3–13 days. The number of dens in simultaneous use is highly variable; Birks & Linn (1982) recorded mink using twice as many dens as that found in Swedish studies.

On the Scottish coast, our 25 telemetered mink used 278 dens over a study period of 6 years (Dunstone, Ireland & Birks in prep.). The characteristics of 233 of these dens were known; the remainder were inaccessible on cliff edges and the like. Overall, 57% were in rabbit burrows, although we recorded a sex difference in their use, males denned more commonly in rabbit burrows. Of the 5% located in human artefacts, two were in stonework (a bridge abutment and a jetty), others were located under discarded corrugated iron and wooden planking or in rubbish tips. Ten of the mink radio-tracked used barns as dens at some time. Although frequently disturbed by farm workers; when full of hay the barns provided a secure, wind-proof den. Some were used by more than one mink, but never simultaneously. Six dens were shared by a male and female during the mating season. Despite the numbers of dens available, particular den sites were favoured by consecutive occupants for reasons that were not immediately obvious.

Many of the dens of coastal mink were located close to a source of fresh water. H. Kruuk (pers. comm.) has suggested that coast-living otters need to rinse the salt from their coats to retain its good insulative properties, and this may also be important for mink.

The dens were located in a wide range of habitats, although rarely on the shore and then only as a temporary shelter used mainly during neap tides. Males spent the winter in dens well away from the shore (Fig. 7.5), whereas the reverse was true of the females. Females moved to dens away from the shore in the spring, but returned to the shore in the summer (Fig. 7.6).

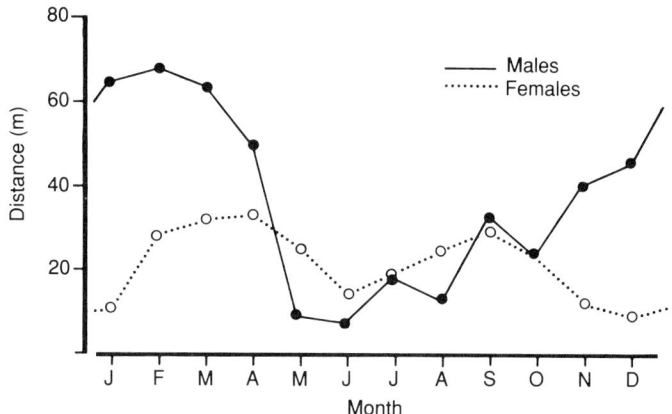

FIGURE 7.5 *Monthly changes in the distance of occupied mink dens from the shore. Source: Ireland (1990).*

PATTERNS OF ACTIVITY

Changes in the level and nature of activity are exhibited by mink in response to seasonal vagaries in the supply of prey and in their own energy requirements and reproductive status. Many species of predators are known to synchronize their periods of activity with those of their principal prey, both on a seasonal and on a daily basis. Most mammals and birds show a characteristic activity pattern governed by the onset and duration of night and day. Three major classes of activity are recognized in mammals with respect to this cycle: nocturnal animals, which are generally only active at night (e.g. badgers); diurnal animals (e.g. squirrels) where daytime activity is usual; and crepuscular animals which make use particularly of the dawn and dusk. Within the major prey groups taken by mink, lagomorphs and small mammals are predominantly active at night, or are crepuscular. Avian prey are mainly diurnal, but this does not preclude their being predated at night. Aquatic prey, on the other hand may be more accessible in daylight because of the limited visual capability of the mink. It is not surprising therefore that the

FIGURE 7.6 *Den of coast-living mink just above the high tide mark. The den is located in the rock in the foreground (top).*

activity patterns shown by such an opportunistic forager as the mink are variable and complex.

Patterns of activity through the 24-h cycle can only be reliably determined by continuous radio-tracking. The resources required to achieve this are considerable and only a small number of studies have been satisfactorily carried out.

Gerell (1969) suggested that male mink generally exhibit a nocturnal pattern of behaviour, although his study did not cover the entire year. For some males a nocturnal behaviour pattern was very marked indeed, but others tended to be arrhythmic. The single female monitored showed a low level of activity when pregnant, but this increased markedly when she had young. Then, both daytime and night-time activity prevailed. A free-living, but recently released, ranch mink continued to show some adherence to ranch feeding times at 07:00–08:00 and 17:00–18:00, as well as night-time peaks in activity.

In England, Birks & Linn (1982) have shown mink in freshwater habitats to be predominantly nocturnal, although one female radio-tracked in January was unusual in that she was active every 2–3 h for about 40 min throughout the day and night. Field voles (*Microtus agrestis*) also exhibit this pattern of activity and it is conceivable that she was synchronizing her rhythm to their activity. Mink engaged in rabbit predation showed little activity for 2–3 days after a kill had been made, when they lay in a burrow and feasted.

In Scotland we have carried out an exhaustive study of the time-budgeting of activity by coast-living mink (Dunstone & Birks 1983, Ireland 1990). A total of 22 769 radio-fixes were collected representing some 3795 h of monitoring of mink 'activity'. In fact, 62% of male and 66.7% of female radio-fixes were from inactive animals in their dens (see Table 7.2). Out of den

TABLE 7.2 *Analysis of all manually collected radio-fixes into each of the four categories of activity and also by sex.*

Activity	Male		Female	
	Fixes	% activity	Fixes	% activity
Inactive in den	3656	62.4	3399	66.7
Active in den	998	17.0	951	18.7
Foraging	790	13.5	585	11.5
Travelling	416	7.1	158	3.1
Total	5860	100.0	5093	100.0

activity was categorized as foraging or travel on the basis of the nature of the radio-signal received. On average, males and females emerged from their dens to forage for slightly less than 3 h per day, although both sexes show reduced foraging activity during late spring and early summer (Fig. 7.7). For the female this reduction may be because the young had just been born in the den. Apart from when they were foraging, males spent approximately 2 h per day travelling, while females only travelled for about half this time. The greatest difference was during the breeding season when males were travelling in search of mating opportunities with the more sedentary females.

Both sexes spent most of their time within dens, where they were mainly inactive. Fig. 7.8 shows the pattern of activity through the 24 hour cycle. Out of den activity is lowest at midday, and highest, at least for males, during the dark. In fact, 84% of male activity occurred during darkness compared with 64% for females.

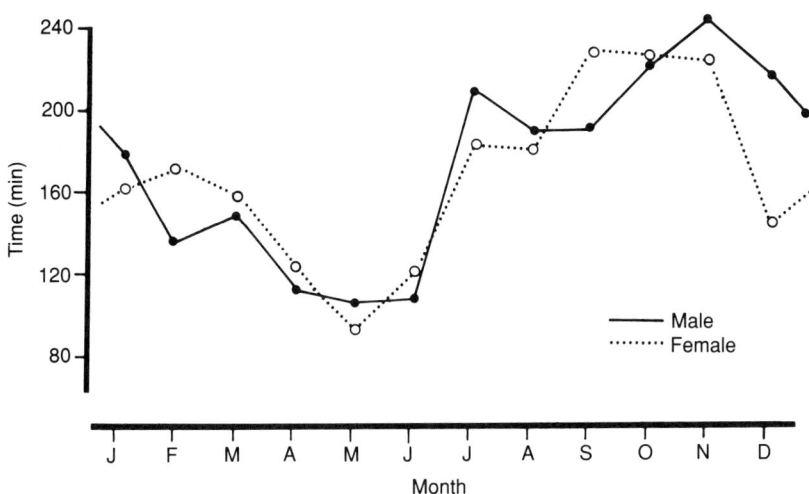

FIGURE 7.7 *Monthly changes in the mean time spent foraging per day. Source: Ireland (1990).*

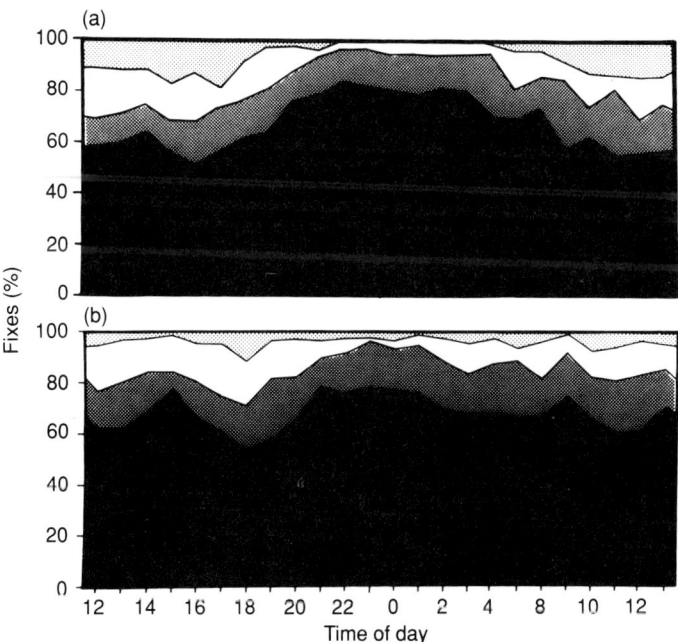

FIGURE 7.8 *Patterns of activity throughout the 24 hours in (a) male and (b) female mink. Source: Dunstone & Birks (1985). Key as Fig. 7.9.*

A greater understanding of patterns of activity can be obtained when they are considered in relation to the mink's annual cycle (Fig. 7.9). During February to April (the rut), the two sexes are active at different times of the day. The activity of males was greatest during dawn and dusk with a lull around midnight. In contrast, females are equally likely to be found active at

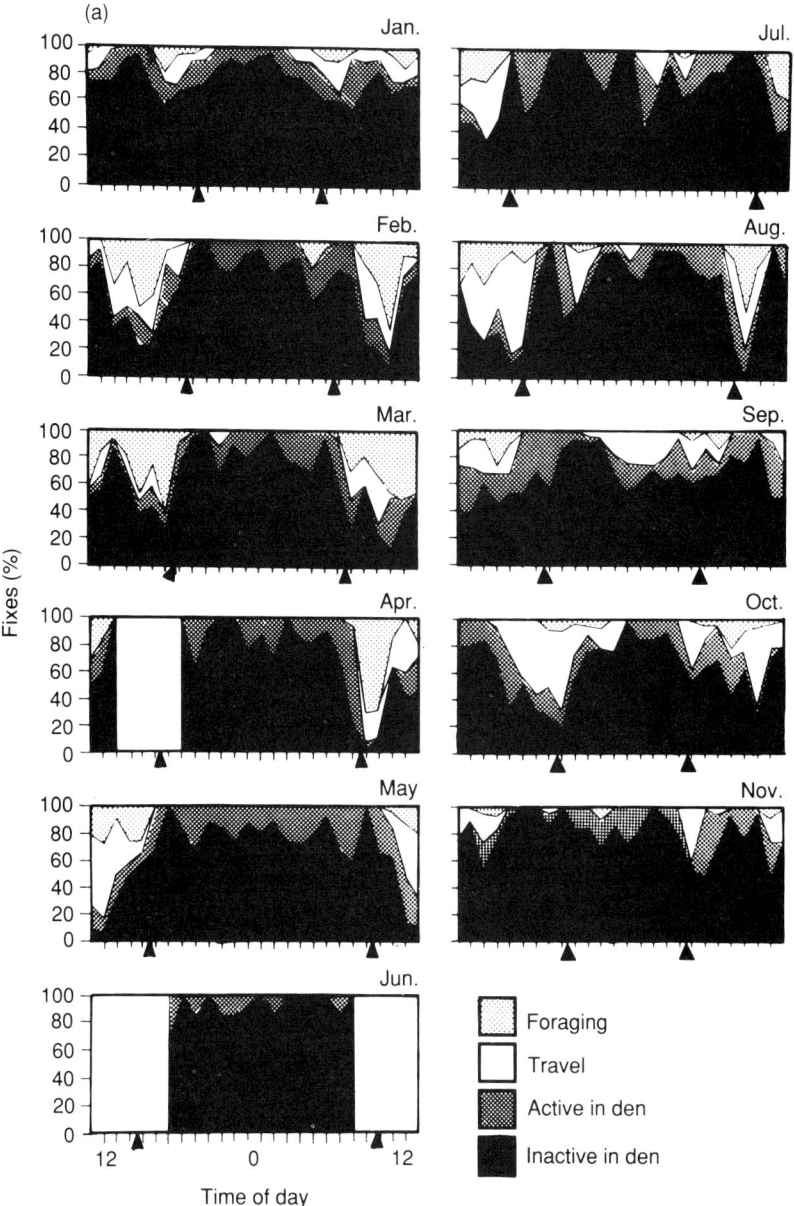

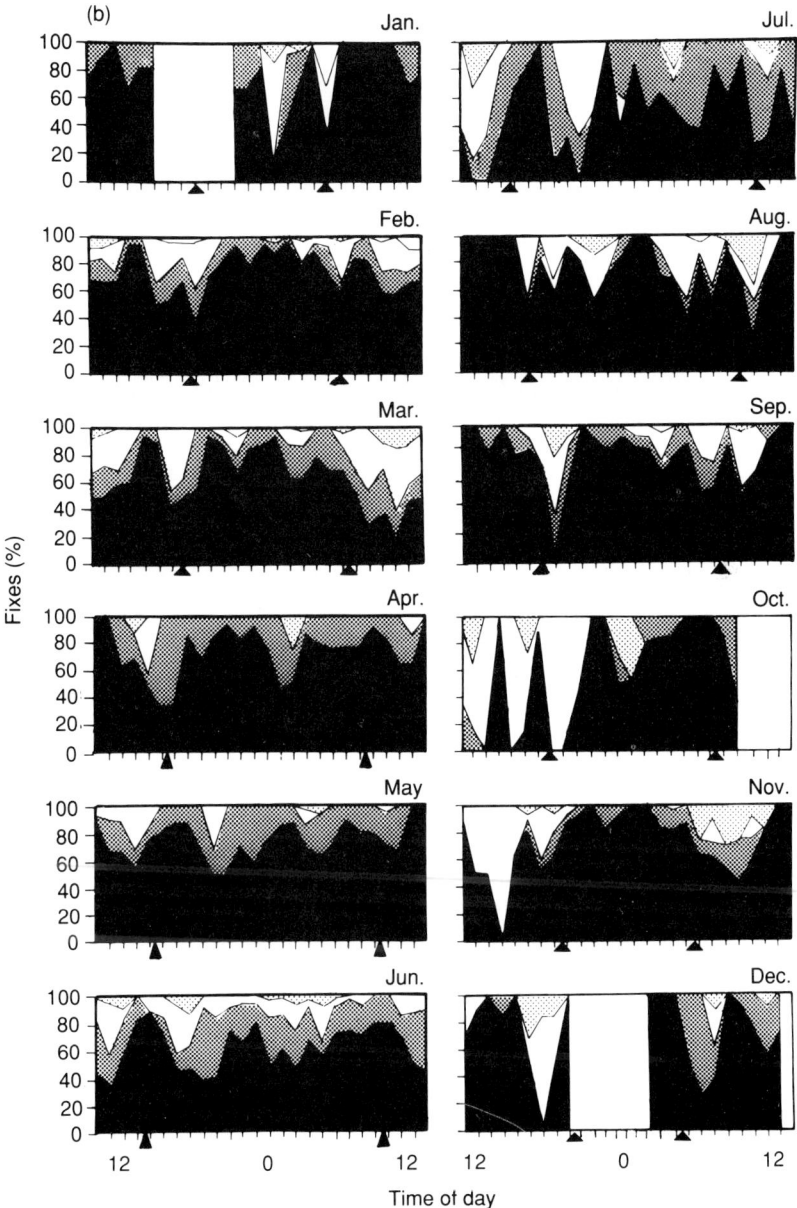

FIGURE 7.9 *Seasonal changes in activity of coast-living mink: (a) males, (b) females. Arrows indicate the time of dawn and dusk.*

any time of the day during these months, although they are only active for some 14% of the time compared to 23% for males. The sexes are also active at different times during May to July, but then activity periods are equivalent for both sexes at about 12%. Females show high levels of diurnality associated with foraging during the demanding process of rearing a litter, while males cram their activity into the short periods of darkness still available. During August to October dispersal occurs, and the males show a similar pattern to that exhibited during the rut with peaks of activity at dusk and dawn and low levels of daytime activity. Female activity is much higher than at any other time during the year. From November to January the activity rhythms of the sexes are indistinguishable. Average levels of activity throughout the night are lower than at other times simply because the nights themselves are longer.

THE DAY TO DAY ACTIVITY OF INDIVIDUALS

There is considerable variation from day to day in both the onset of activity bouts and their duration. This is best illustrated with reference to data collected by remote monitoring of animals that used particular dens for extended periods. This method of recording is particularly valuable in aiding our knowledge of the behaviour of breeding females before and after weaning, when they are restricted to one den. Such a record for a female is shown in Fig. 7.10. During this observation period she was feeding mainly on the shore, usually from 05:00 to 06:00 and from 17:00 to 18:00. A male, monitored using the same technique, was quite consistent in the time he returned to his den, usually in the hour after dawn, following an activity bout in the early morning (Fig. 7.11). His foraging usually began about 50 min after dusk.

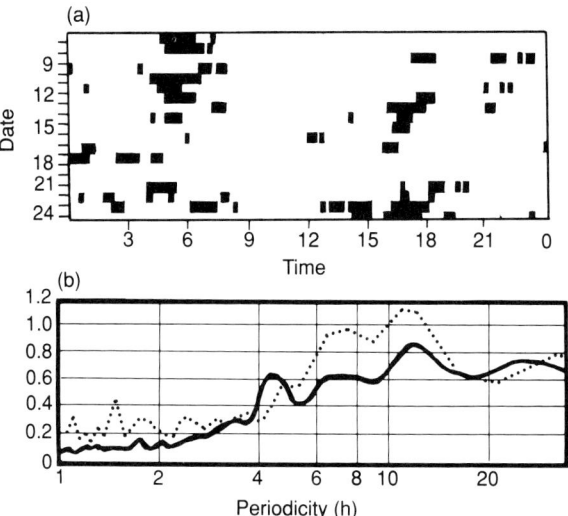

FIGURE 7.10 *(a) Activity diagram of a female mink when rearing young in June (data collected using remote monitoring apparatus), (b) periodogram of the same data, peaks in the graph indicate periodicity in activity. Source: Ireland (1990).*

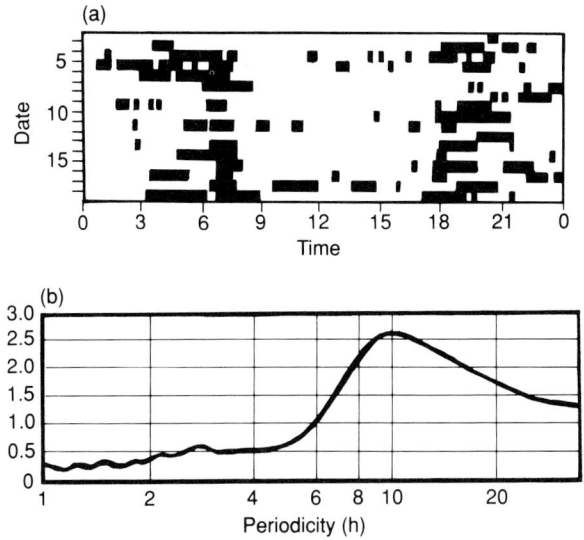

FIGURE 7.11 (a) Activity diagram of a male mink in November (data collected using remote monitoring apparatus), (b) periodogram of the same data, peaks in the graph indicate periodicity in activity. Source: Ireland (1990).

PERIODICITY IN BEHAVIOUR

A periodogram is a graph depicting the likelihood of an event occurring at a specific time. Such analyses provide a convenient and informative way of depicting cyclical events in activity data. Tall, narrow peaks on the graphs indicate rigidly occurring cycles of activity. Examples of these are shown in Fig. 7.10b. Here, for a female in the breeding season, a cycle of activity was noted to occur at 12-h intervals, with a subsidiary peak every 7 h. Outside of the breeding season a female may show activity every 6 h. A similar analysis for males (Fig. 7.11b) showed less frequent bouts of activity with a periodicity of about 10 h.

ACTIVITY LEVELS THROUGHOUT THE YEAR

Despite their considerably larger size compared to females, male mink spend a similar amount of time foraging. The maintenance requirements of mink, as estimated from laboratory studies, predict a 1.5 times greater energy need for the male. Assuming both sexes to forage equally efficiently and to take similar sized prey, this would require male mink to forage for appreciably longer each day. The fact that males do not forage for longer than females suggests either that they are hunting more successfully or that they are taking a prey of greater utility. We have seen that males prefer lagomorphs and larger prey types generally. Thus the increased size of the male appears to result in an increase in foraging efficiency by virtue of their increased success in catching larger prey.

The occurrence of diurnal activity of mink may be explained, at least in part, by the short gut transit time, that is the speed at which the mink digests its meal. In any one meal a mink can only eat sufficient to fill its stomach, and then must wait some 3 h before it can eat again. For a mink eating rabbit prey secure within a burrow this presents no problem, but for animals foraging on small terrestrial or aquatic prey this will require the animal to frequently leave its den in search of food. This will be manifested as short bouts of activity with the mink returning to the security of the den in the interim. Since only one or two such bouts may be possible during darkness in high summer, the mink will need to be active partly during daylight hours if it is to obtain adequate food. This situation will be exacerbated for the breeding female by the added demands of providing for the kits.

OTHER EFFECTS OF THE PHYSICAL ENVIRONMENT

Weather
It has been suggested that male mink may increase their level of activity in cold weather. In one case a cessation of activity seemed to be related to the lowering of temperature. This occurred in conjunction with high humidity causing ice to form on vegetation. The influence of rain and wind is not particularly marked. In extreme weather conditions the activity of ecologists is probably more limited than that of the mink!

Time and Tide
In mink, as with other species which inhabit coastal regions, there is good cause to suspect that the tidal cycle (with a period of 12.4 h) may also exert an influence on activity. The effect of the tidal cycle on activity patterns is difficult to discern in an opportunistic forager such as the mink which might suddenly abandon the shore and switch its predation to terrestrial prey. However, when mink were foraging on the shore females showed a distinct preference for activity when the tide was low (Fig. 7.12). A slight preference for incoming tides was noted. This pattern was not shown by males, who were as likely to be found on the shore at any state of the tide. The bulk of the mink's aquatic prey (rock-pool inhabiting fishes and crabs) is found in the littoral zone (between low and high water) of the shore. The limited diving capability of the mink suggests that most prey should be taken when the tide is out. Diurnal foraging for aquatic prey might arise, as has been already suggested, if moderately high levels of illumination are required for underwater hunting. However, neither sex has yet been shown to adhere to daylight for aquatic hunting.

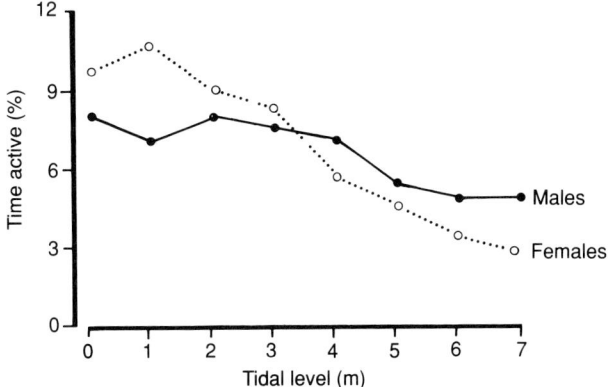

FIGURE 7.12 *The effect of tide on the activity of coast living mink.*
Source: Ireland (1990).

ACTIVITY PATTERN OF THE EUROPEAN MINK

According to the interpretation of the Russian literature by Youngman (1982), this species is mostly crepuscular, with the main periods of activity from 16:00–22:00 and before dawn (05:00–10:00). In very cold weather European mink are reputed not to leave their dens.

METHODS OF DETERMINING MOVEMENTS, HOME RANGE AND ACTIVITY

To study movement patterns, the ability to identify individual animals is necessary. Fortunately this is facilitated in mink by the possession of individually distinct patches of white fur on the chin, chest and belly (see Plate 9 and Fig. 2.2). Mink can be ear-tagged using small, numbered metal (Monel) tags. The tag needs to be fitted well down the base of the ear to prevent it from being pulled out. These are occasionally lost, and therefore the ventral spot pattern is also recorded.

Live trapping can fulfil a number of roles:

(i) collection of population status data: e.g. population density, sex ratio, proportion of juveniles in the population, residency, information on the arrival of new individuals in the study area and home range estimation,
(ii) measurements of weight, size and age,
(iii) attachment of transmitter packages.

TRAPPING TECHNIQUE

To provide information on movements and to assess the extent of the home-range, mink must be live-trapped and then preferably radio-tracked for some considerable period. Live traps are also widely used in control operations, but in addition Fenn Mk. IV break-back traps are becoming more commonplace in the UK. Although wooden treadle live-traps have been used in Wisconsin (McCabe 1949), nowadays traps are usually constructed of galvanized wire ('Weldmesh'). Chicken wire is inadequate, as mink are quite capable of biting their way out! The traps measure about $60 \times 20 \times 12$ cm. Particular attention has to be paid to the door locking mechanism to prevent corrosion and hence jamming. In a coastal site this can become a considerable problem and a stainless steel mechanism may be advisable.

The traps are usually baited, although the necessity of using flesh as an attractant is debatable, since the inquisitive nature of the mink towards any novel objects in its range can lead to many captures. Probably the best 'bait' is the odour of another mink, particularly during the rutting season. The provision of bait will not diminish trapping success, and if a population is to be repeatedly trapped, bait may provide an incentive for a wary animal to enter a trap. It also provides a meal for an animal whose foraging excursion has been severely interrupted by capture. Repeated captures on successive nights can seriously interfere with the well-being of some trap-prone individuals, and it may be advisable to temporarily suspend trapping activities if this happens. The type of bait used will depend on availability. During control operations, fish or rabbit flesh have commonly been used. As far as possible the bait should be fresh although mink will occasionally be attracted to putrefying material. During scientific study it may be advantageous to use bait not normally encountered in the diet (for example, dead day-old domestic chicks or white laboratory rats). This is particularly important if diet is also being investigated, otherwise the results can be confounded. Indeed, this was a problem in some of the early dietary studies conducted on mink in the UK (Day & Linn 1972). Since fish was occasionally used as bait, its occurrence in the scats was viewed with suspicion.

The location of the trap is all important, particularly in view of the mink's predilection for staying in the vicinity of the waters edge. However, traps should not be sited too close to the water in case of flood or high tide. Traps are usually positioned facing downstream, presumably because animals are more likely to use an overland route when moving up river instead of swimming against a current. Always attempt to make use of features in the habitat that channel a mink's movement to a highly specific path, for example by siting them under a bridge. Obstacles to movement, such as stone walls, can also make suitable locations for traps. Since this will ensure that the mink will encounter, although not necessarily enter, the trap.

It is normal to disguise the trap, but the extent to which this is really necessary has yet to be systematically investigated. During control operations, bare traps are frequently deposited in appropriate locations. Most people prefer to disguise the traps since this also prevents interference from passers-by, and in winter, can provide all important insulation from the elements. A variety of materials will be available at the trap location and should be used to blend the trap into the background and to reduce the neophobia (fear of novel objects) that the mink might otherwise show. Suitable materials include tussocks of grass, branches, hay and rocks. It is particularly important to camouflage the entrance to the trap.

Since mink occupy linear home ranges, it is not usually appropriate to lay out a trapping grid. Rather it is more effective to employ a line of traps along a water-way (river, coast or lake margin) making use of habitat features and natural runways that the 'mink-thinking' trapper can visualize! Sites of regular deposition of faeces and the occurrence of dens can give a clue to areas commonly used. Equal spacing of traps is difficult to achieve, but ideally one trap every 200 m of good habitat should prove effective. Successful traps can be left *in situ* and unsuccessful ones moved to new locations.

In scientific studies it is important to reduce stress to the mink by visiting all traps as early as possible during the morning. This procedure should of course also be followed by those undertaking control operations. Unfortunately this does not always seem to be the case, and on many occasions mink have been found starved, or drowned in cage traps after a flood.

PROCEDURE FOR HANDLING MINK

Recaptured mink can often be identified within the trap on the basis of their ear-tag or ventral spot pattern. If, however, prolonged investigation is required it is appropriate to anaesthetize the mink. This is best achieved by transferring the trapped animal to a perspex-sided box. A dry mixture of air/ether is passed through the box until the mink is unconscious, giving anaesthesia for 2–4 min (Fig. 7.13). This is just adequate (usually!) to attach

FIGURE 7.13 *A trapped mink being prepared for examination.*

an ear tag, sketch the ventral spot pattern, sex, age, measure and weigh the mink. The mink is then allowed 10 min recovery before being released at the point of capture. Longer periods of anaesthesia, such as may be required for attachment of a radio-collar, may be achieved by intramuscular injection of ketamine hydrochloride (trade name 'Vetalar', Parke Davis).

It is possible to age mink accurately using a variety of laboratory techniques (see Chapter 9); however, none of these are suitable for use in the field with live animals. Approximate ageing can be achieved in the field by palpating the baculum (penis bone) of males. Since the width of this bone increases with age it is usually possible for an experienced recorder to ascertain whether he is dealing with a sub-adult or adult male. Differential growth of the bones of the skull leads to subtly different facial characteristics which can also help decide whether an individual is a juvenile or adult. In the case of females the presence of white hairs on the back of the neck which are derived from healed mating bite wounds indicate that she is older than 10 months (mink are born in May and are mated for the first time in the following March). These criteria only allow adults and juveniles to be distinguished. Tooth wear has occasionally been used to age mink in the field. Care has to be taken, however, because individuals from some populations, can have badly worn teeth as a result of eating Crustacea. Mink known to be 3 years or older can often be distinguished by the presence of grey hairs in the coat. The most accurate method of determining the population age structure is to capture and mark mink as kits.

Sex and reproductive status can be assessed in the field. The testes of male mink remain within the abdomen except during the breeding season when they descend into the scrotum. When the female is receptive, the vulva may evert. The presence of fresh mating wounds and scabs on the neck are an indication that mating has taken place. The foetuses can be felt by gently palpating the abdomen of pregnant females, and the presence of obvious nipples will indicate that birth has taken place and that lactation is occurring. Tentative information on the litter size can also be obtained by counting the number of active nipples (see Chapter 8). The general body condition can be assessed on the basis of the extent of the inguinal fat store and body weight.

TRAPPING SUCCESS

Trapping success can be highly variable, and is particularly dependent on seasonal factors, especially reproduction. Control operations can also affect the success of a trapping campaign since mink can become very wary. Comparisons of trapping success between the numerous studies that have been conducted are difficult since there are many variables, for example, trapping intensity, the initial population size, and perhaps most importantly, the capability of the trapper. Data can be expressed as the percentage of captures per trap-night, (where one trap-night is one trap set for 24 h). Using this measure, Ritcey & Edwards (1956) recorded 93 caputres in 556 trap-nights along 1.2 miles of stream during late summer trapping between

1951–1954. In more extensive year-long studies, trapping success has varied over the year. Mink are more frequently captured in late summer, when juveniles predominate, and during the mating season, when adult males prevail (Gerell 1971). This seasonal pattern seems to be very common and correlates with an increased mobility of the mink at these times. In Scotland (Dunstone & Birks 1983, Ireland 1990) captured 82 mink 431 times in 4412 trap-nights, distributed as shown in Fig. 7.14. The rut, in particular, is an easy time to catch mink. Success in trapping males increased from mid February to a peak in March.

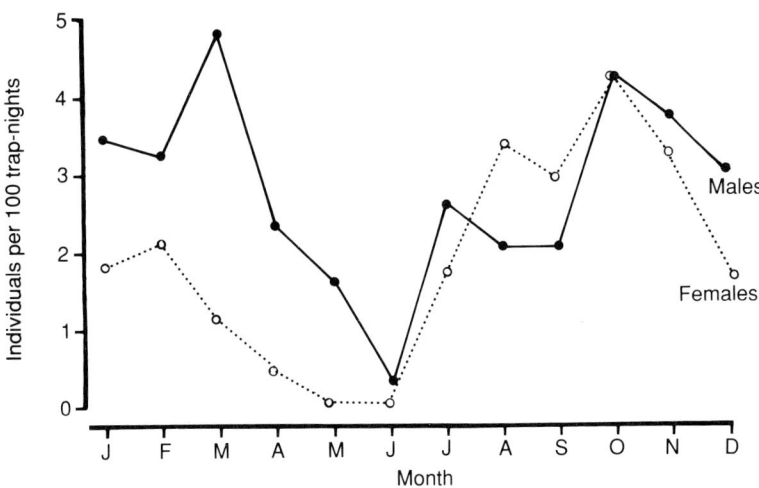

FIGURE 7.14 *Monthly fluctuations in trapping success during live trapping operations. Source: Ireland (1990).*

For control trapping operations, which involve permanent removal of mink, there is usually a high frequency of initial captures which falls off quite quickly (Fig. 7.15), and all trappable resident mink will have probably been encountered within 1 or 2 weeks. Subsequent captures will tend to be of animals that have extended their ranges to encompass land previously occupied by these mink. Individual mink are relatively easy to capture once, but the proportion recaptured varies with sex, age and status. Juvenile males may show a high recapture frequency because they are subjected to a low territorial pressure, whereas the low recapture rate of adult males may be indicative of a transient population. Juvenile females are considerably more difficult to recapture than adult females, perhaps indicating a high dispersal frequency. Adult females are always the most sedentary part of the population. Trapping success is at its lowest during May and June for both sexes.

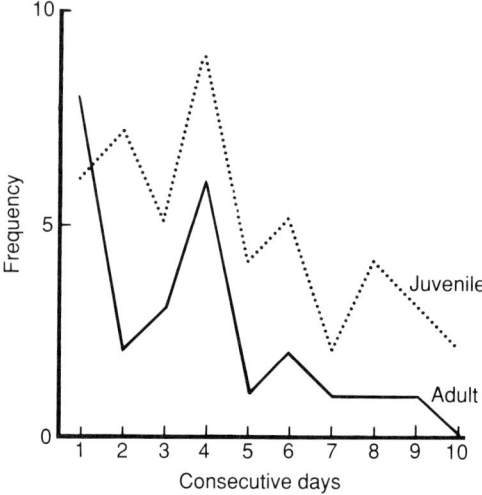

FIGURE 7.15 *Frequency distribution of initial capture on the river Rönnea, southern Sweden. Source: Gerell (1971).*

Trapping studies can provide useful information on the status of the population and the condition of individuals. However, as a tool for determining the movements of animals it is very limited. Once trapped an animal's movement is curtailed until release and unlikely to be resumed with its original motivation.

RADIO-TRACKING TECHNIQUE

To reduce the interference with an animal's lifestyle, it is better to monitor it's activity by remote means. This involves the attachment of a radio-transmitter onto the mink whose movements are then followed using a radio receiver and antenna. Three methods are available for attaching the radio-package. It can be made into a collar or a harness, or for short-term studies, it can be glued onto the fur of the back. Back packs are not generally successful since they can cause skin abrasion and restriction of movement (e.g. Birks & Linn 1972). Care has also to be taken with the attachment of radio-collars. The best results have been obtained using transmitters with a long whip antenna woven into the leather fabric of the collar. The alternative is to use a tuned loop antenna where the brass aerial also forms the necklet. In America, free-ranging mink have been telemetered following implantation of the transmitter into the abdominal cavity (Melquist & Hornocker 1982). In the UK, welfare legislation has prevented this technique being used in the past.

The radio-devices transmit at a rate of between 30 and 40 pulses min^{-1} at a chosen frequency. In the UK, most telemetry studies are conducted using the

Photos by Nigel Dunstone unless otherwise credited.

![Mink carrying perch to den.]

1. *Mink carrying perch to den.*

2. *Mink carrying woodmouse.*

3. *Mink investigating horse clams.*
Photo: Dave Hatler.

4. Mink in alert posture. 5. Young mink sitting on rock den.

6. Mink peering at fish prey.

7. *Adult mink in alert posture.*

8. *Prey size selection experiment using shore crabs.*

9. *Ventral spot pattern used in identifying individual mink.*

10. *Typical riverside habitat favoured by mink.*

11. *Mink habitat in northern Spain.*

12. *Coastal habitat utilized by mink in southern Scotland.*

13. Pastel coloured mink.

14. Tree root den.
15. Breeding den with accumulation of scats.

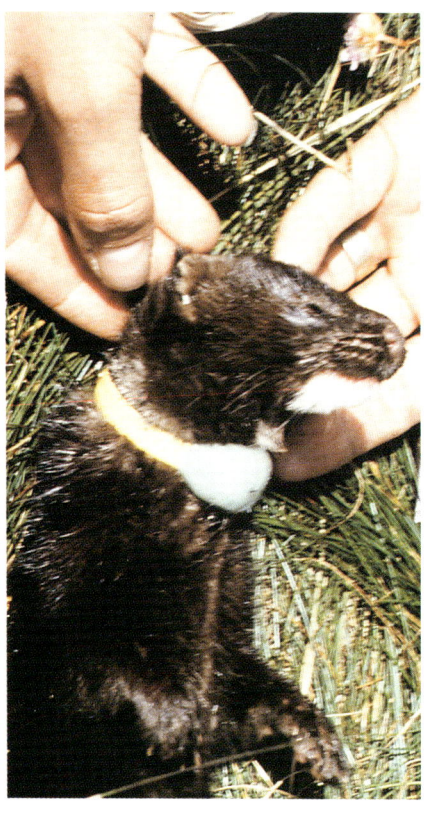

16. *Radio-collared mink.*

17. *Taking measurements from an anaesthetized wild mink.*

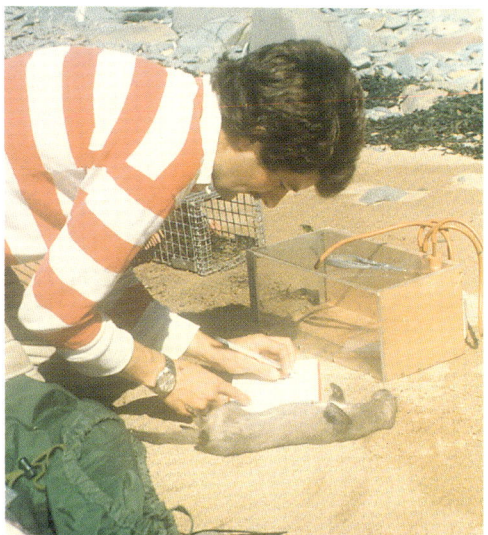

18. *Barn used by mink as a winter den.*

19. *"Asking for trouble?"* Free-range chickens on a mink infested river in North Yorkshire.

21. *Young mink kit at approximately six weeks.* PHOTO: DAVE HATLER.

20. *Gamekeepers gibbet with mink, stoat and weasel.* PHOTO: ROB STRACHAN.

22. *Newly born mink kit.* PHOTO: DAVE HATLER.

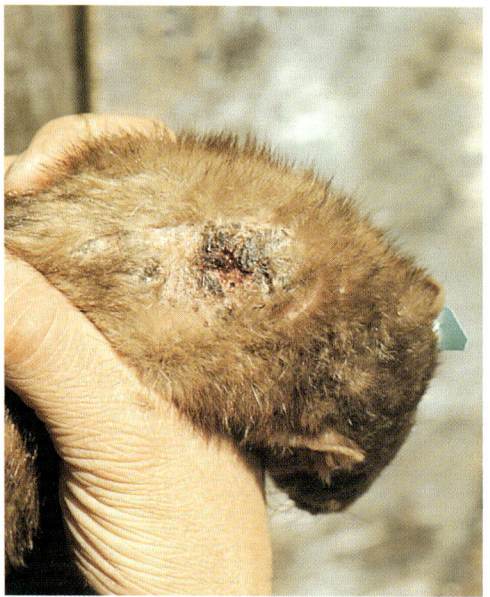

23. *Severe wound on the neck of a female, a result of a mating fight.* PHOTO: DAVE HATLER.

24. *Pattern of moult in a coastal mink from Vancouver Island.* PHOTO: DAVE HATLER.

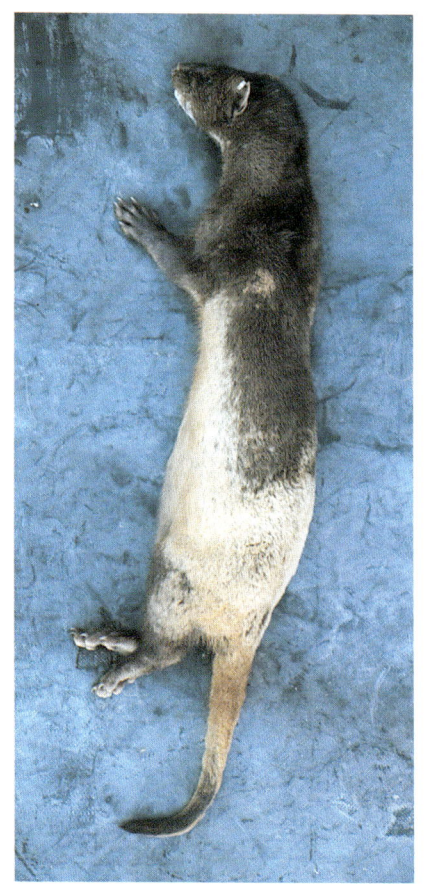

173 MHz bandwidth, whereas in North America the 102 MHz bandwidth is commonly used. Each transmitter is tuned to emit at a particular frequency. Since frequencies as close as 20 Hz can be discriminated by the receiver, a number of mink can be simultaneously tracked in the same study area. The maximum life of the radio-collar is dependent on power output and battery size. No animal, of any species, should be required to carry a transmitter package greater than 5% of body-weight and, preferably it should be considerably less. For mink, the typical weight of a transmitter package is 20–30 g which would normally give a maximum battery life of between 8 weeks and 9 months. Transmitters rarely achieve this maximum possible life however, since eventually water enters the package as it becomes worn away by abrasion with the substrate. Birks & Linn (1972) attempted to determine whether attachment of a transmitter deleteriously affected a mink's behaviour before embarking on their radio-tracking study. There is a period of inactivity after release which probably results from the stress of capture, handling and anaesthesia. Wild radio-tagged mink were commonly observed to scratch at their collars. This behaviour was noted for 72 h in a captive animal, after which her behaviour was apparently normal. Of a sample of eight mink examined, only three showed any decrease in weight which might indicate interference with feeding. It may be expected that wearing a collar is likely to interfere with aquatic foraging. However, the incidence of aquatic prey in the diet was not influenced in these animals.

Radio-tracking a small nocturnal, frequently fossorial (underground living) carnivore rarely permits direct observation even though the operator may be only a few metres away from the subject. Under these conditions much has to be inferred from the quality of the signal received from the mink's transmitter. In ideal situations, with a direct line of sight, preferably from an elevated position, the transmitter can have a range of 1 km. The mink's closeness to the ground, its subterranean habits (e.g. in rabbit burrows or rock dens) usually reduce the effective range to 50–100 m. Fortunately, the mink occupies a linear habitat; hence a walk along the waterway is usually adequate to locate the animal. The topography of the habitat has a marked effect on the ease with which the radio-signal can be detected. Generally, following the signal until it does not get any louder ultimately leads to the den. Gross changes in signal strength occurred when the mink was active out of its den. Characteristic changes in the signal can give information on whether the mink is swimming or diving. In the latter case the signal disappears, only to re-appear when the mink surfaces.

Small 'Beta lights' (MacDonald 1978) which emit light can be fitted to a collar. These small devices are so effective that a mink fitted with a Beta light may be more visible at night than a mink foraging in the open during the day. Their effect on foraging behaviour is not known, but presumably they are equally visible to the prey.

Dunstone & Birks (1985) found they could reliably categorize mink activity into four classes depending on the nature of the radio-signal received from the animal's transmitter.

Inactive in den The mink remained in one location and the radio-signal did not fluctuate, that is, the animal was probably resting.

Active in den The mink remained in one place, but the signal strength fluctuated constantly, that is, the animal was probably eating prey or grooming.

Foraging The mink was in the open and moving slowly with wide fluctuations in signal strength.

Travelling The mink was out of its den and moving through its home range at a speed which precluded searching for food.

The out-of-den activities were occasionally supported by sightings of the animal.

Since mink spend most of the 24-h cycle within a den, it is useful to monitor their activity remotely from within these refuges. This technique can only be used to monitor their presence/absence at a given den, and hence the duration of out-of-den activity. The nature of this activity cannot be determined. The technique simply involves connecting the telemetry receiver to a data-logger, which switches on, records the signal, either electronically or on a chart recorder, and then turns itself off at regular intervals (Fig. 7.16). The signal

FIGURE 7.16 *Automated data recording from telemetered mink in a den. Photo: Nigel Dunstone.*

from the mink's transmitter is recorded if it is present at the den. The technique can be extremely useful for recording from breeding females while they are restricted to using one den by the presence of the kits. However, most mink use a number of dens and this severely limits the quality of data that could otherwise be obtained by this cost-effective technique.

In some studies it is adequate to obtain once or twice daily radio-fixes of a mink's location, particularly when the object is to determine the extent of the home range. However, there is a great likelihood that signals will be received only from within dens, and the home range calculated may not fully indicate the area used during foraging and territorial patrol. A number of researchers have analysed for how long the animal should be monitored to reveal the extent of the home range. Birks & Linn (1982) consider that 80% of a mink's home range is revealed within 5 days. Tracking periods greater than 10 days are thus desirable since the estimate of home range size increases with the number of active fixes recorded up until that time. For one female mink, only 65% of her final range area was demonstrated in the first 27 active fixes recorded (the average number of active fixes collected per month). However, this represented 100% of the linear length of her eventual range. Thus, the duration of a radio-tracking study can be crucial in determining the accuracy of the home-range measurement.

To monitor adequately the rather unpredictable behaviour of mink requires considerable effort and determination; particularly during spells of inclement weather, when the mink may remain safely secure within a rabbit burrow for 3 days without emerging! Ideally the animal should be tracked throughout the 24-h cycle by a team working in shifts. Only occasionally has this been possible (e.g. Dunstone & Birks 1985, Ireland 1990). More generally the scientist has to split the 24-h record over a number of days with a consequent loss of information.

RANGING BEHAVIOUR

HOME RANGE OR TERRITORY?

Animals tend to exhibit characteristic spacing patterns. The area over which an animal moves during the course of its activities is referred to as its home range (Burt 1943, Jewell 1966). Hence, home range is defined by the fact that an animal is using a particular area, not by how it uses it. The home range must, by definition, satisfy the animal's biological requirements. Therefore, we would expect large animals to inhabit large home ranges simply because of their great metabolic requirements. The relationship is complicated because large and small mammals do not feed on prey of the same size, and certain assumptions must be made about the relative abundances of their prey. This relationship between home range size and body-weight has been explored by McNab (1963) and Gittleman & Harvey (1982). As might be expected, animals which have to 'hunt' for their prey require large ranges. The home ranges of carnivores and those herbivores that actively search for their widely distributed food are, on average, four times larger than the home ranges of browsers that 'crop' their food.

It is frequently noted that some animals, the mink included, actively defend part or whole of the home range against some other members of the species, for part or all of the year. This defended area is referred to as a territory. Territorial defence is costly for both the time and the energy that needs to be expended. There may also be a risk of injury. Given these considerations we must assume that the benefits of being territorial must outweigh these not inconsiderable costs. Various reasons have been forwarded for the evolution of territoriality. For mink, a guaranteed access to a food supply, access to receptive females, and defence of a den and young are considered to be important.

Territoriality may evolve when two major conditions prevail. Firstly, there is competition among individuals: a requirement for this to occur is that the resource is limiting. Secondly, the resource that is being contested, for example food, must be economically defendable. The defended area should be just large enough to provide access to the resource. Thus, we might expect territory size to change over the yearly cycle as prey availability fluctuates and the demands of reproduction become apparent.

SEX AND SPACING PATTERNS

The variety of spacing patterns exhibited by members of the family Mustelidae have been examined by Powell (1979) in an attempt to explain the variation in spacing patterns that was becoming evident from field studies. Powell considered the basic mustelid spacing pattern as one of intrasexual territoriality, with males having larger territories, overlapping those of females. This was well documented for weasels, stoats and American pine martens. Evidently mink adhere to this schema. River otters showed a variation on intrasexual territoriality, while sea otters and European badgers did not exhibit this type of spacing, probably because of their colonial behaviour.

HABITAT INFLUENCES ON HOME RANGE

It is common for mink researchers to measure the extent of the home range as the linear length of the watercourse the animal traverses, even though the animal may occasionally spend extended periods of time away from water. Nevertheless, for most of the year this is probably a fair assumption to make. For riverine habitats it is generally assumed that both banks are available for use, a fact usually borne out by radio-tracking studies.

Much of the activity of male and female coast-living mink occurs within 100 m of the high water mark during most months of the year. This also seems to be the case for mink living along rivers. Here it is generally found that home ranges are basically linear with some extensions away from the main river along feeder streams or to areas where there are high densities of terrestrial prey, such as rabbits.

Since food is one of the main determinants of home range size we would expect the area over which a mink may roam to be related to the productivity of that habitat. The marked sexual dimorphism of body size suggests that male mink have greater energy demands and consequently larger home range sizes than females. The availability of potential den sites and stability of the social system will also be of importance. Accordingly we must examine these parameters for each of the habitat types we investigate.

Keynote studies contributing to our knowledge of the home range behaviour and movements of mink along rivers and lakes are those conducted in the USA by Melquist *et al.* (1981), in Sweden by Gerell (1970), and in England by Birks (1981).

Riverine habitats
Birks (1981) used telemetric and trapping techniques to compare mink movements on two contrasting rivers: the oligotrophic River Teign and the relatively eutrophic River Exe. The Teign is an acid river rising from the granite mass of Dartmoor and supports a salmonid dominated fish fauna. Within the study area the river was bounded by steep-sided deciduous woodland containing good populations of rabbits. The River Exe, in contrast, supports populations of coarse fish and some waterfowl, with a low density of rabbits. The most remarkable aspect of this latter study area was its situation, partly within the city limits of Exeter, a thriving market town with a population of 100 000. The surrounding habitat consisted of residential areas, gardens and recreational areas with footpaths bordering the river. The disturbance level from man and dogs was high, but the mink were afforded some privacy by the fringe of emergent vegetation, principally alders (*Alnus glutinosa*) and willows (*Salix* spp.).

The home range characteristics of mink studied in a variety of habitats are given in Table 7.3. The extent of the home range exhibited by three of the nine mink radio-tracked by Birks are illustrated in Fig. 7.17 (Birks & Linn 1982). The mean home range length for a male mink inhabiting a stable social environment on the Teign was 2.79 km, while a female ranged over 2.87 km. During the rut (February) on the more eutrophic Exe, two males used 2.9 km of river each, and a female 0.5 km.

In Sweden, the study area, in the Valley of Fyledalen, was located principally along the Klingavälsan, a stream that meanders through pasture land bordered by alder and beech trees (Gerell 1970). Here adult male mink roamed over 2.5 km of stream each, while juvenile males tracked during November and December occupied between 1- and 2-km ranges. Considerable restriction of range was noted in a pregnant female mink tracked during April and May, but outside the breeding season female home ranges were only slightly smaller than those of males.

TABLE 7.3 *Mean home range lengths occupied by adult male and female mink in England and Sweden.*

	Mean length (km)	Range
Rivers		
Males (England)[b,c]	2.50	1.6–4.4
(Sweden)[a]	2.64	1.8–5.0
Females (England)	2.10	1.2–3.2
(Sweden)[a]	1.85	1.0–2.8
Lakes		
Males (England)[b,c]	1.90	
Females (England)[b,c]	1.46	
Coast		
Males (Scotland)[d]	1.50	
Females (Scotland)[d]	1.09	

Source: Gerell (1970)[a], Chanin (1976)[b], Birks & Linn (1982)[c], Dunstone & Birks (1985)[d].

Lacustrine habitats

Birks also studied the movements of a male and a female mink at Slapton Ley, a eutrophic lake with an abundance of prey, particularly waterfowl and coarse fish, and with dense cover along the margins (Birks & Linn 1982). Here the male and female mink radio-tracked ranged over considerably smaller areas, 1.46 and 1.9 km of lake margin, respectively. In Sweden, Gerell (1970) radio-tracked a male mink along a 5-km strip of lake shore bordered by extensive reed beds and willow which was also rich in waterfowl and fish prey. The mink population in this area was very low which might explain the large range occupied by this animal.

Coastal habitats

Only two studies have been conducted on mink movements in a coastal habitat. In British Colombia, Hatler (1976) studied a population on Vancouver Island using trapping techniques and direct observation to establish a density of one male mink per 0.74 km of coastline; female density was higher at one animal per 0.44 km.

In Scotland (Dunstone & Birks 1983, Ireland 1990) the ranging behaviour of 44 mink was investigated during a 6-year study. These mink achieved a

FIGURE 7.17 *(a) Home range of a male mink on the River Teign, Devon. Letters mark the locations of dens. Dens A and B are located in rabbit burrows 225 m from the river. (b) Home range of a male mink in a suburban region of the River Exe, Devon. Letters mark the locations of dens. (Dens A to D are rabbit burrows). (c) Home range of female mink at Slapton Ley (a lake). Letters mark the locations of dens around the periphery of an area of carr, marsh and reedbed. Source: Birks & Linn (1982).*

Ranging Behaviour

(a)
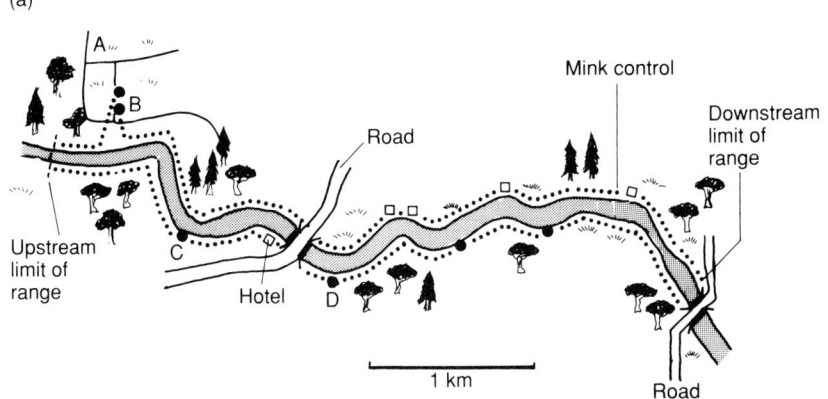

(b)
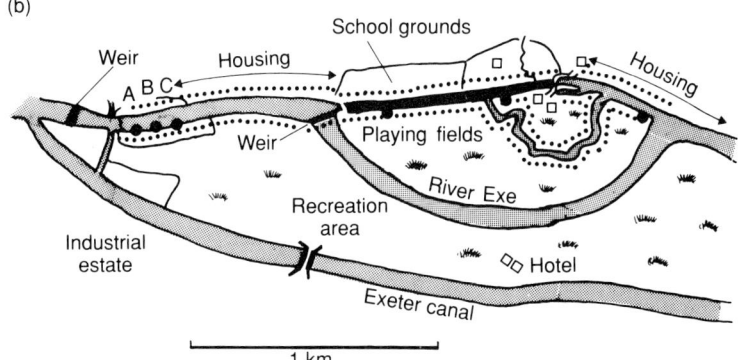

(c)
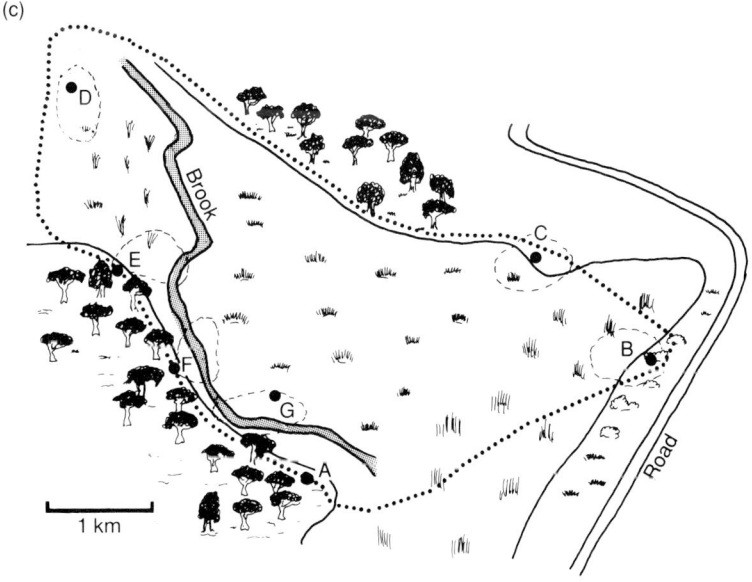

greater density due to individuals occupying quite small home ranges, with a greater than usual degree of intersexual overlap; the mean length of coast occupied was 2.65 km for males and 1.24 km for females. Assuming the mink do not stray far from the high water mark, this represents an area of land amounting to 12.3 ha for males and 4.9 ha for females.

SEASONAL CHANGES

Our Scottish study provided the only complete record of mink activity throughout the year, and revealed changes in the extent of the mink's range according to its seasonal requirements (Fig. 7.18) (Dunstone & Birks 1983, Ireland 1990). Male home ranges were markedly greater than those of females from February to June. Outside the mating and breeding season, sex differences were neglible in most months. Male home ranges varied from a mean of 5 km in March, the peak of the rut, to a minimum of 1.5 km in June. Those of females ranged from a minimum 0.3 km in May, when lactating, to just over 2 km in August.

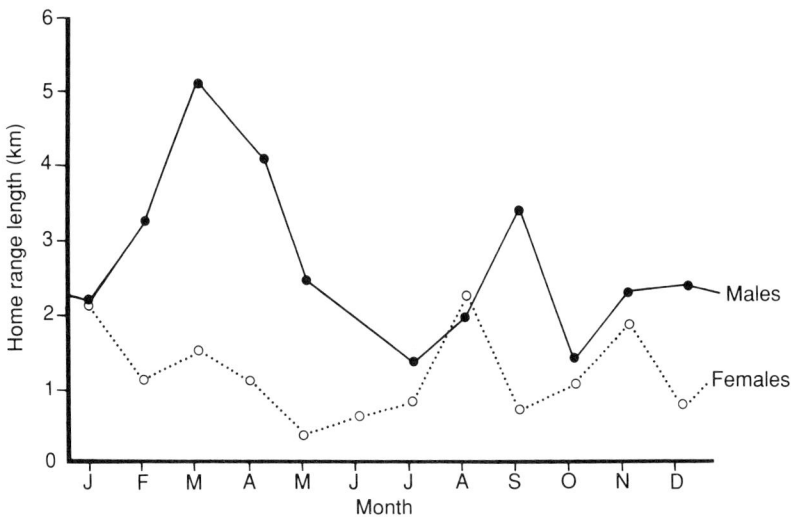

FIGURE 7.18 *Monthly changes in the extent of the home range for male and female coast-living mink. Source: Ireland (1990).*

TERRITORIALITY IN A LINEAR RANGE

Despite these fluctuations, mink are considered to have a less variable range size than similar species that occupy a non-linear range. The time and energy costs of patrolling a linear territory are very high. Chanin (1976) has suggested that linear territories compare unfavourably with the more conventional two-dimensional ones that are characteristic of the weasel or polecat in terms of the economics of defence against intruders. The area of interaction with intruders is at the extremity of the territory. To increase the foraging area encompassed by the linear home range by four times would also engender an increase in border patrol by four times. For the two-dimensional territory, however, assuming that territory shape is unchanged, a four-fold increase in territory size would only involve a doubling of the border to be patrolled. This arises because the increase in area is proportional to the square of the increase in circumference. Birks & Linn (1982) suggest that this constraint may explain the low variability of mink home ranges.

Since this constraint will be more pronounced on the larger male who has to defend a larger territory, this may also explain why males are more likely than females to abandon the linear territory system in favour of foraging away from watercourses.

RESIDENCY AND FLUX

In some studies there appears to be a high population turn-over. Dunstone & Ireland (1989) recorded an average residency of only 8 months (see Fig. 7.19), although one female was resident for 3 years. The greatest time a male was observed resident in the area was 25 months. There was a large population of transient individuals passing through the study area, particularly adult and sub-adult males during the rut.

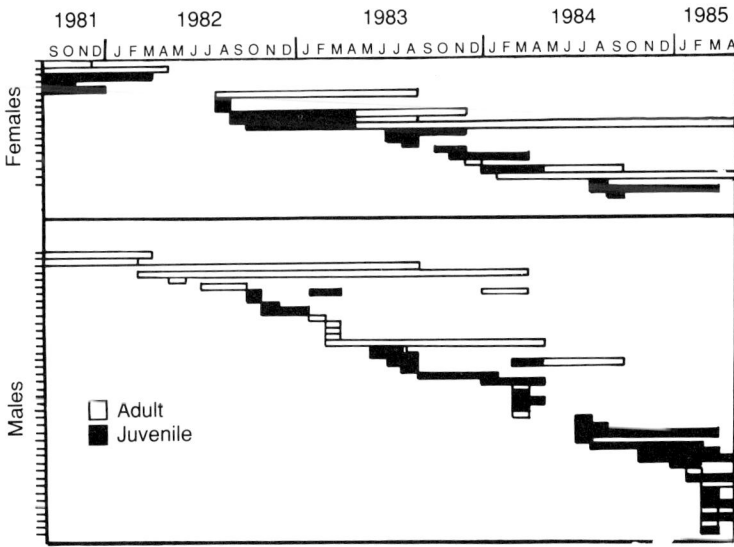

FIGURE 7.19 *Trends in residency exhibited by coast living mink. Source: Dunstone & Ireland (1989).*

Transient individuals

An understanding of the behaviour of this section of the mink population is very important in many respects. Their wide-ranging movements are indicative of animals searching for an area in which to settle and establish a territory. As we have seen, most transient individuals are captured during the rut as sexually mature males, or less commonly during the early autumn when juvenile animals predominate.

The presence of transient mink also has to be considered in the context of control operations. Since it is probably from these populations that mink killed by man will be replaced. Colonization of new areas is also likely to be by transient and dispersing mink.

Social instability

Mink home ranges are found to be longer in areas of social instability. The main cause of instability in the pattern of territories observed is the death or removal of one of the territory holders. The cause of this is often persecution of the mink by trapping, shooting and hunting with hounds. Generally, there is an increase in length of the territory occupied by contiguous animals. The increase in length comes about when a neighbouring animal is removed allowing an unchallenged extension of range. In such conditions on the River Teign, three males were observed to extend their ranges to over 5 km, almost twice the length of that occupied by a male in a stable area. This also happens in un-persecuted populations when one of the animals dies from natural causes. Ireland (1990) recorded two instances when, as a result of mink deaths, adjacent territory holders expanded their ranges. In one case (see Fig. 7.20a,b) a juvenile male took up residence adjacent to a resident male which subsequently died of unknown causes in October. By November of that year the juvenile male had expanded his range to cover most of the territory of the

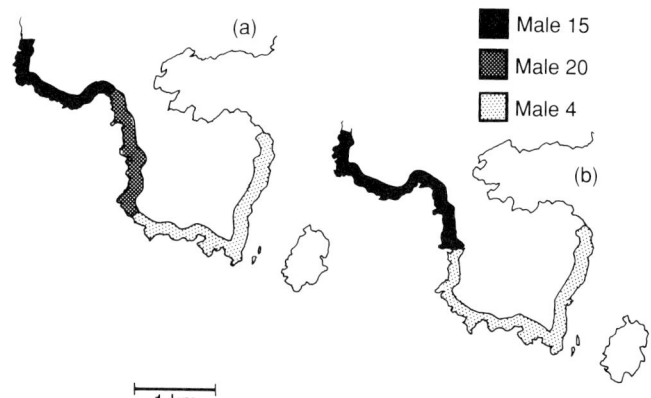

FIGURE 7.20 *The change in male home ranges which occurred after the death of a male 20: (a) November 1984; (b) February 1985. Source: Ireland (1990).*

dead mink. This is of course indicative of a territorial system, where the range of an individual is limited by anothers intolerance of its presence, enforced by overt aggression or olfactory warnings. As soon as one animal is removed, range incursions occur. Thus, after the removal of a competitor a mink will range over a longer territory than it requires to provide for its sustenance. If and when a new individual arrives, the territory may be reduced again as the surplus land is uneconomic to defend.

EUROPEAN MINK

Studies in Russia (Danilov & Tumanov 1976) suggest that the density of the European mink in the Leningrad region may reach some two to three animals per 10 km of shoreline, while in the Pskov region some seven to twelve animals may inhabit the same length of shore. Males occupy an area of some 32 ha, while females occupy smaller areas of about 26 ha. Of eight territories examined, the largest encompassed 3 km of riverbank with a mean of 2.4 km.

MOVEMENTS WITHIN THE HOME RANGE

Intensity of use

There is a great irregularity of use of different areas within the home range. This may arise from the distribution of food and the spatial distribution of suitable dens. On the basis of radio-tracking observations it is possible to identify the so-called core areas where mink prefer to forage. Birks & Linn (1982) consider that such areas occupy about 9% of the home range, yet mink spend approximately half of their active time there. On their River Teign study area, half of these areas were associated with rabbit predation. Gerell (1970) noted a similar phenomenon for mink inhabiting the Valley of Fyledalen. From Fig. 7.21 it can be seen that the male used the central parts of

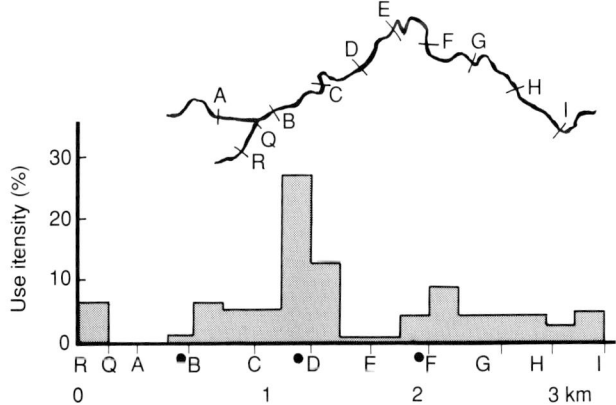

FIGURE 7.21 *Home range use intensity by an adult male mink.*
Source: Gerell (1970).

his range most intensively, an area of good cover provided by reed beds. During December he was more commonly located in an area of young oak plantation and presumed to be feeding on rabbits, approximately 300 m from the stream.

Thus core areas are those where dominant prey abound, and may change seasonally according to the availability of particular prey species. Activity outside the core areas frequently consists of patrolling and scent marking of territory rather than foraging.

Daily pattern of movement
Gerell (1970) found that when a mink used a core area its activity pattern was characterized by to and fro movements probably not exceeding 300 m in extent. Eventually the mink will visit another area and repeat the pattern of restricted movement. At the same time, the movements formed a pattern of larger oscillations with a similar appearance, but extending over the entire home range. This is shown diagrammatically in Fig. 7.22, and can be interpreted as territorial patrol and intensive use of particular foraging areas.

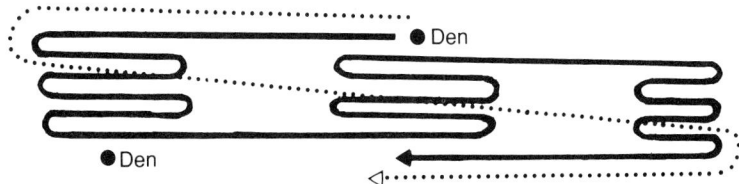

FIGURE 7.22 *Diagrammatic representation of the movement pattern of a mink. Solid line represents small scale patrolling movements which combine to eventually cover the entire home range. Source: Gerell (1970).*

Speed of movement
The speed of movement of adult male mink varies depending on the type of activity being performed. Gerell (1970) found daily oscillatory movements within the home range to be about $0.4\,\mathrm{km\,h^{-1}}$. Passage between two core areas was more rapid, between 4.7 and $2.8\,\mathrm{km\,h^{-1}}$.

Patterns of den use and movements between dens
The types of dens occupied by mink have already been discussed. Their number will be limited by the number of potential den sites available. It is appropriate to discuss here the movement pattern between these dens. Birks & Linn (1982) found mink to use from between two and ten dens. The mean number used between November and April was four, while between May and October nine dens per animal were in use. In their study, the mean distance between dens was approximately 500 m, with females possessing a density of dens almost twice that of males. The majority of den stays lasted less than 1

day (see Fig. 7.23). A similar pattern has been found in Sweden (Gerell 1970). In Scotland also, telemetered mink changed dens almost on a daily basis (Ireland 1990). Males changed dens most frequently during the rut, while lactating females were the most stable occupants.

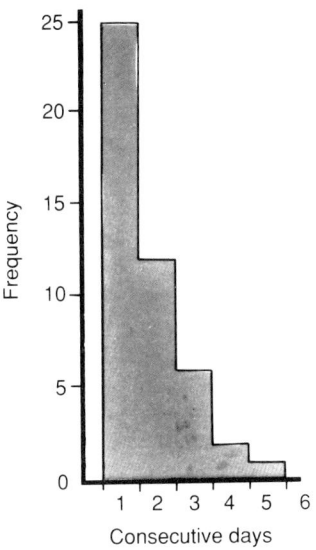

FIGURE 7.23 *Frequency distribution of the duration of den occupancy. Source: Birks & Linn (1982).*

EXTENSIVE MOVEMENTS AND DISPERSAL

Mink usually leave the territories occupied by their mother during August, at which time they are about 3 months old. Rarely an animal may 'inherit' a territory on the death of its mother, and even less frequently it may annexe off part of her territory for its own use while she is still alive. Of 40–50 young mink raised to independence during a 2-year study in Devon, only five animals established territories within the study area (Birks 1989); the remainder dispersed from their natal area. One was run over by a vehicle 10 km from its place of birth. In Sweden even greater distances of travel have been recorded, one mink moving 40 km (Gerell 1970).

SOCIAL INTERACTIONS

THE DEFENCE OF TERRITORY

It is never straightforward to establish that an animal is territorial. Rarely is overt aggression observed directly and the pattern of scent deposition that an animal uses to mark its territory is far beyond the capability of the human nose to discern. However, if we examine the pattern of adjacent home ranges throughout a particular habitat then some interesting conclusions can be drawn. Figure 7.24 shows the distribution of male and female territories in the two commonly used habitat types. In each case, it is apparent that each mink occupies a discrete home range in which other individuals, at least of the same sex, do not settle. The oscillatory patrol movements also point to the existence of a territorial system.

Generally the males do not tolerate other males within their range. Females may tolerate their daughters for part of the year of their birth. A female may evict an overlapping male from her territory particularly when she has kits. As we have seen, she tends to occupy a smaller range at this time and will be

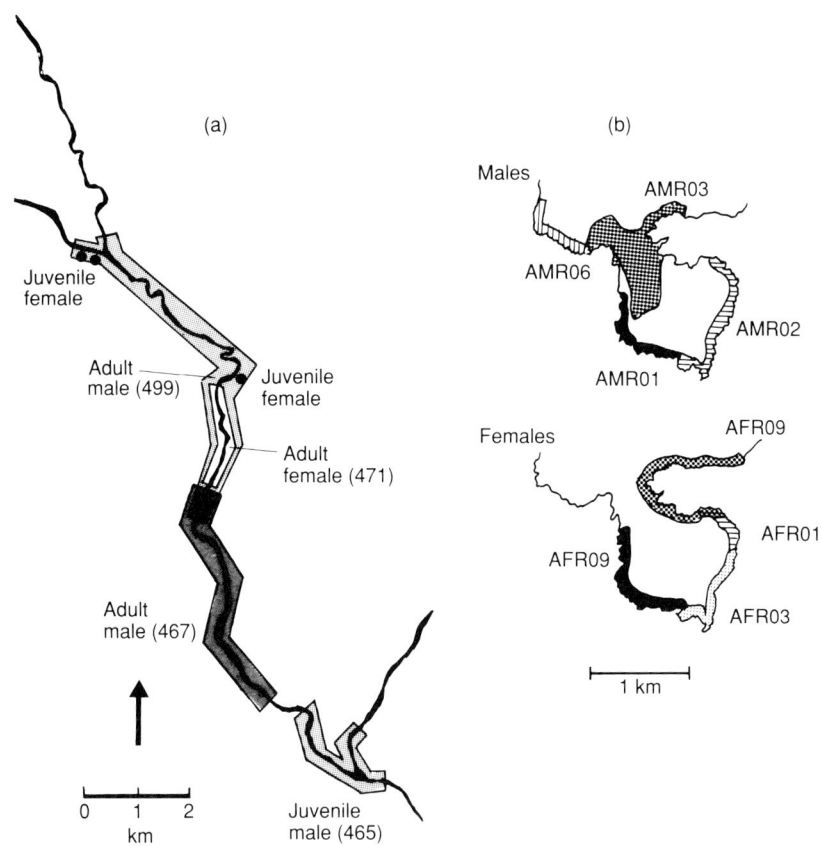

FIGURE 7.24 *(a) Spatial distribution of mink in the valley of Fyledalen, Sweden in December 1969. Source: Gerell (1970). (b) Spatial distribution of male and female mink on the Ross peninsula during 1982–1983. Source: Dunstone & Birks (1985).*

placing higher demands on it. This may explain her intolerance of competition.

INTERSEXUAL INTERACTIONS

It has already been suggested that overlap may exist between male and female territories, whereby a male's territory may include a part or whole of one or more female ranges. This has been substantiated by several trapping studies (e.g. Gerell 1970, Chanin 1976), but the frequency and extent of this phenomenon is difficult to establish. Birks (1981) found only a third of females to have extensive overlap with males. However, the size of the mink's range (and probably that of it's defended territory) may vary seasonally, so this relationship is probably constantly changing.

We estimated that there was evidence of range overlap between males and females for 29 of the 43 months (67%) of our coastal study (Dunstone & Birks 1983, Ireland 1990). For those months encompassing the rut—December–April—the overlap was 95%. In June, a female passed within 1 m of a den occupied by one male although no interaction took place. In the same month he was trapped close to one of her breeding dens. In July, he was observed stealing small mammal prey from another breeding female on the Mull of Ross. In October 1981 simultaneous tracking of a different male and female showed these animals to occupy contiguous ranges, but in December this male had enlarged his range to include that of the female (Ireland 1990).

Male/female associations during the breeding season

A female in heat attracts a number of males to her vicinity. Traps located close to a female den during March are almost always successful. In one case, four traps caught a total of six mink, with 10 captures in 10 trap nights. Traps in which females had been captured were occasionally completely excavated by males trying to get to them. On one occasion, a male tried to drag a trap containing a female into his den!

Females did not always have males in attendance at their dens during the rut. Males were recorded at female dens in February where they may have been waiting for the female to return from foraging or to emerge, or were themselves resting after copulation. Ireland (1990) records one male to overcome a female in a struggle that lasted only 5 min, after which he ran off. He returned 40 min later, just as she was re-entering her den. Copulation then ensued for 13 min before he dragged her into her den where they remained for 11 h.

INTRASEXUAL INTERACTIONS

While intrasexual overlap is rare, home ranges are not completely exclusive. There is often spatial, if not temporal, overlap at the borders of adjacent ranges. Levels of overlap between males were about 10% outside of the breeding season. Most of this probably occurs during the absence of one of the territory holders. This suggests that perhaps only specific areas of the home range are defended as a territory. During the rut, spatial overlap between male ranges can rise to 55%. Few studies have attempted to simultaneously radio-track mink in adjacent territories to determine whether there is temporal avoidance.

Radio-tracking can reveal some interesting relationships and interactions between individuals (Ireland 1990). In November, one male, Fred, weighing 960 g took over part of the range of an adult male known to have died from natural causes the previous month. The remainder of the dead male's range was occupied by yet another juvenile male who weighed only 910 g. The adjacent range, encompassing the area known as Fox Craig, was occupied by a larger male, Arthur, weighing 1140 g. By December Fred, now weighing 1030 g, was mainly using inland areas of forestry, although he was also trapped three times on the east side of the peninsula. During this month Arthur had expanded his range along the eastern side of the peninsula which he continued to do through January with occasional visits to the rabbit warrens on the steep cliffs of Slack Heugh (see Fig. 7.1). Fred was trapped for the last time in February; his range now extended along the west and south sides of the peninsula, and as far as Slack Heugh in the east. On one foraging trip Fred entered Slack Heugh while Arthur was present within a den. At the beginning of March, Fred was trapped on the west side of Brighouse Bay by which time he had put on a lot of weight (1370 g). He was never trapped again and is presumed to have left the area. This example is indicative of range overlap at the boundaries of contiguous territories. A similar situation was noted for females. In February 1984 a sub-adult female, Jasmine, was trapped intermittently for 9 days. At this time she had fresh scabs on the back of her neck suggesting that she had been recently mated. She appeared to have a stable range centred on a barn and extending as far as Ross Bay. The only other female know to be in the vicinity at this time was an adult female (Shula) mainly occupying Ross Island, but who also made foraging trips to patches of forestry on the mainland. By March, Jasmine was resident in Ross Bay, and was trapped three times within the mainland part of Shula's territory. The system of home ranges of sub-adult females appears to be in a state of flux between September and February each year. In September 1982, Marianne was trapped at Fox Craig. In October the same trap captured Shula. Marianne was seen swimming from Ross Island and attacked Shula on her release from the trap. Marianne screamed submissively and swam across to Ross Island. The following month both animals had moved away from this area.

Apparent range incursions in December did not usually lead to the displacement of the resident individuals, but by late January the resident male population had started to change. In January 1983 and 1984 the previously resident sub-adult males both left. In one case an adult male (M4) weighing 1170 g encroached on the range of a smaller sub-adult (M10), who subsequently left. Between the end of January 1983 to early February 1985, four sub-adult males, previously resident outside the main study area, were captured indicating the considerable mobility of these animals at this time.

Resident male mink did not leave the study areas at the same time as transient (usually sub-adult) animals arrived. Trapping indicated that many other males were present temporarily or passing through the same stretch of coast during March. During one night a male was tracked for 8.7 h during which time he traversed 8.6 km of the peninsula (see Fig. 7.25). On his travels he provoked screaming from two males caught in live-traps and he passed through three female home ranges during his ramblings. In one female home range he approached her den but was attacked by a male who was in attendance. He left immediately, but approached the den again twice more that night.

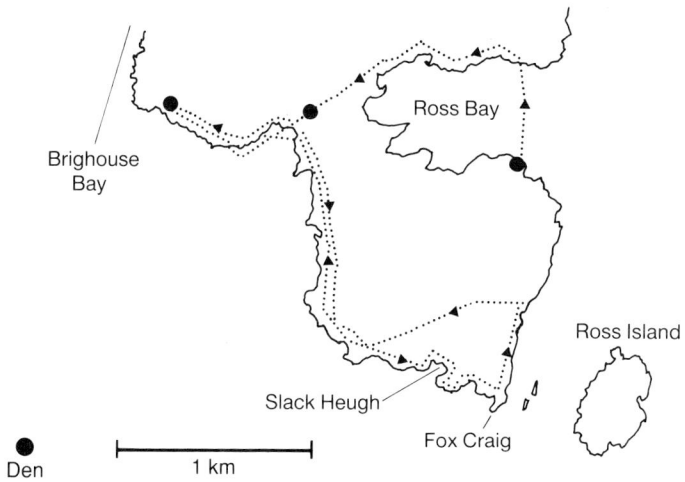

FIGURE 7.25 *Path followed by an adult male during the night of 26–27 March 1984. Source: Ireland (1990).*

OVERT AGGRESSION

Few species of animals inflict serious injury on one another as a result of intraspecific aggression. Fighting among mink is well known and can be a very violent affair. Long & Howard (1987) confirm that mortal combat can occur in this species:

> ... two mink began fighting, tumbling around and thrashing noisily in the weeds of the dry marsh. They emitted no vocal sounds, ... but continued struggling for about five minutes. By then the larger mink seized in its jaws the head of the smaller mink and held it for approximately 30–45 seconds, as the struggle ceased. Then the large mink seized in its mouth the thorax of the smaller for 10–15 seconds. Finally the large mink slinked away, without further attention to the now dead smaller mink ... The skull was found to be pierced (puncture of the frontal bone), and the thorax was also punctured.

Such direct observations of agonistic encounters between mink are rare

FIGURE 7.26 *Aggressive postures of mink.*

and have usually only been documented from laboratory investigations (Fig. 7.26). They are most commonly witnessed during the rut between males in competition for mates. In Scotland Ireland (1990) observed that shortly after mating with a female, a sub-adult male left her den and was involved in four short fights with two other sub-adult males. Later that evening while in the vicinity of her den, he was engaged in another fight and was presumably beaten, since he moved off at high speed and was never trapped in her range again. Other males trapped during the rut often had scabs under the fur around the cheeks, and nape of the neck indicating they had been involved in fights.

THE SIGNIFICANCE OF SCENT MARKING

In mammals, the deposition of scent can have various functions. It can be used to discriminate between individuals and to provide an indication of reproductive status and condition; in addition, it may deter predators or warn conspecifics (Macdonald 1985). In the context of the territorial system it may allow an animal to orientate itself within its territory and indicate when it has reached a boundary. The scent of a territory owner may deter territorial intruders. After encountering the scent of a territory holder an intruder will be able to recognize the owner when challenged. This may allow them to avoid escalated fights, during which serious injury is likely to be incurred (Gosling 1986). Such a system is relatively cheat-proof since it takes a long time to scent mark a territory effectively, and hence only the resident territory owner achieves this.

Territorial scent marking in mink may be achieved by use of glandular secretions, particularly those of the anal gland, and deposition of scats (see Chapter 5). It is not possible to estimate the importance of the former in wild populations since deposition sites are not obvious to man's grossly inferior nose. Scats probably also have an olfactory signal function, they are also easily located visually and can be mapped. Male mink fed to surplus deposit up to 10 scats per day, females about six, and they can persist for a month. They are likely to be found in the same location from month to month, indicating a sign-post function (Robinson 1987), but are distributed more or less evenly throughout the territory with no obvious concentration at boundaries between adjacent animals' ranges.

The violent reaction of a mink to the scent of an intruder leaves one in no doubt about the signal value of a scat. Birks (1981) placed a fresh scat from a territorial female at the regular scatting site of the male occupying the neighbouring territory. On discovering the introduced scat the male sniffed it for about 15 s and showed considerable signs of excitement including bottle-brushing (pilo-erection) of the tail. The male then defaecated beside the introduced scat, producing a greenish jelly-like material.

CHAPTER 8

Sex and Society

One of the advantages of working with mink is that many aspects of its biology have received the detailed attention of physiologists. This is particularly the case for reproduction. Mink exhibit many specialized adaptations of their mating system and reproductive physiology, making them a fascinating subject for study. A definitive study of the reproductive processes involving the breeding biology of ranch mink has been carried out by Enders (1952), and more recently Sundqvist *et al.* (1988) has extensively reviewed the voluminous literature regarding hormonal control of reproduction.

Female mustelids exhibit three basic types of gestation pattern. The first type involves relatively short gestations and immediate implantation of the blastocyst following fertilization. This pattern is typified by the polecat (*Mustela putorius*) with a gestation period of 40–42 days and the weasel (*Mustela nivalis*), with gestation lasting 35–36 days. The second group, which includes the mink and the striped skunk (*Mephitis mephitis*), have a variable gestation period, with females fertilized early in the season having the longest gestation. The final group exhibit long gestations following a prolonged period of delay before the fertilized egg implants in the uterus; the badger (*Meles meles*) and the stoat (*Mustela erminea*) are classic examples.

*A*NATOMY OF THE REPRODUCTIVE SYSTEM

MALE

The reproductive system of a male consists of the testes, epididymes, vas deferens, prostate gland and penis. The testis, which is egg shaped, is composed of a large number of tubes, the lining of which produces sperm. Once produced, the sperm move to the epididymis, a highly coiled tube which lies on the surface of the testis, where they mature. A fine tube called the vas deferens transports the sperm to the penis. The prostate gland, which produces fluid to give volume to the ejaculate, empties its contents into the urethra at the base of the penis. The penis of the mink is approximately

5–6 cm long, and lies enclosed in a sheath on the belly. Within the penis is a hook-shaped bone (see Fig. 8.1), the os penis or baculum, which may serve to position the penis up against the cervix during ejaculation ensuring that the sperm are deposited as close to the cervical canal as possible.

FIGURE 8.1   *The baculum.*

FEMALE

In anticipation of the forthcoming breeding season the ovaries, oviducts and uterus increase in size in the spring. It is difficult to establish the reproductive state of a female from external examination since the vulva swells only slightly during heat (oestrus). It is likely that the male will be aware of her physiological readiness for mating from the olfactory information available.

SEASONAL BREEDING

Mink breed once a year, with both males and females becoming fertile in their first year. During the mating season a female usually comes into heat two to three times. In northern Europe, Russia and North America the mating season lasts for approximately 4 weeks starting in early March.

The reproductive cycle of the mink, like the fur-growth cycle, is under the controlling influence of hormones, modulated by changes in day length. Testicular development occurs in the autumn and is stimulated by increasing day length.

A small gland at the base of the brain, the pituitary, is stimulated to secrete a hormone (gonadotrophin) which acts on the ovaries to stimulate egg formation in the female. The pituitary gland produces another hormone which promotes the growth of the reproductive tract in both sexes and initiates glandular activity in the testis and ovary. In the female this results in the release of oestrogen and in the male testosterone. These 'sex hormones' initiate mating behaviour.

MATING BEHAVIOUR

When in the vicinity of a female in heat, a male usually demonstrates his 'interest' by emitting a characteristic chuckling noise which seems to serve a 'contact' function. Initially he will examine areas that have been scent-marked by the female, frequently by overmarking with his own glandular secretions and urine. Then courtship starts in earnest, usually with a violent fight. If the female is in heat she will bend her tail to one side. The male grasps the female by the neck, and treading on her hips attempts to straighten her arched back to allow coitus to occur. Hatler (1976) describes the mating behaviour of a coast-living Californian mink:

> ... I watched as a known, tagged male dragged her up to a small pocket among the rocks. He had a hold on the back of her neck, but she was squealing and struggling violently. For the first five minutes she struggled frequently, but she did so only occasionally as time went on.
>
> ... It was clear that each time the female struggled the male tightened his grip on her neck; at times he visibly chewed. He maintained the neck-hold for the duration of the mating, which was 1 hour 50 minutes. Both animals showed signs of fatigue afterward; the male, in fact, had semi-paralysis of the hindquarters, and shortly after releasing the female he collapsed when he tried to jump onto a nearby rock where she had gone.

Such lengthy copulations are not unusual; Enders (1952) noted the longest he observed in ranch mink to last for 3 h. However, much shorter periods *in copulo* have also been observed. Despite the stress that the female endures there are numerous observations of females soliciting further the attentions of males, as Hatler (1976) records:

> Commercial fisherman, D. Arnet, was working on his boat and, hearing squealing down the dock, went to investigate, He found a copulating pair of mink on the floor of a nearby open boat. The male had the female by the neck, and she was squealing and struggling. Arnet checked on them frequently, and about an hour after the first observation he found they were missing. Leaning into the boat for a closer look, he found that they had moved up under the seat and his presence interrupted them. The female jumped out of the boat, ran up the dock, and went into the woods 50–75 m away, but the male stayed under the back seat. Approximately 20 minutes later the female returned, jumped back into the boat and went to the male, and they started again!

The male can inflict severe damage on the female during copulation, particularly within the confines of a cage from which she cannot escape. Incidents are reported in which males have killed females, death resulting from the canines penetrating the base of her skull. In northern England I have trapped wild female mink with severely infected neck wounds inflicted during mating.

Multiple matings are the norm, and these need not necessarily be by the same male. Hatler (1976) similarly recorded the mating season as a time of severe stress for female mink:

> I visited a small island on my study areas to check on a tagged female there. I spotted her, alone, but seconds later found her in the grip of a big, tagged male from the adjacent mainland. He released her in less than a minute, and his subsequent behaviour left me in little doubt that he had mated her just prior to my arrival. The female, upon her release, ran towards her den but was intercepted by another male.

After a scuffle, accompanied by much squealing, the female broke away and escaped to the den. In the two hours which followed this (juvenile) male persisted in his attempt to catch the female and almost succeeded in one other squealing, biting encounter. He was still there when I left. The next day I again saw a mink on the island, and I saw mink (presumably males) swimming to this island on two of the following three days. On 5th June I again visited the area and found the female curled up as though asleep, in a spot where I had seen her sleeping before, but she was dead.

Autopsy showed three punctures and numerous bruises on the back of her neck, but these were not bad enough to cause death. Her digestive tract was empty and she had no fat. Hatler (1976) believes she died of exhaustion directly as a result of harassment by the males.

TRANSIENT MALES

Most studies involving the trapping of mink show an elevated capture rate of new males in March. These males have been termed transients (Birks 1981) as they often pass through the study area within a short time. Resident males have also been observed to leave their territories and assume transient status. Most of these transient individuals are sub-adults. In Scotland about half of the population of sub-adult males adopted the transient strategy. Transient adults and sub-adults were occasionally observed to remain in the vicinity of oestrus females, and may have achieved matings.

As the end of the mating season approaches, fewer and fewer receptive females will be available, and there will be intense competition between males for those still in heat. Males will be forced to move increasingly further away from their normal ranges in search of mating opportunities. Ireland (1990) estimated that there will be three or four males competing for each receptive female at this time.

OVULATION AND FERTILIZATION

Unlike animals that ovulate spontaneously when their eggs are mature, mink are induced ovulators. If the follicles have reached a certain size the stimulation of mating causes a surge in their growth resulting in their release from the ovary (ovulation) within 36–42 h. The stimulation of the vagina by the penis is thought to be responsible for the initiation of ovulation. There is some evidence that intromission need not necessarily occur, and that in some cases the rough fighting foreplay of courtship may lead to ovulation.

During intromission nerve impulses travel from the vagina to a control centre in the brain, the hypothalamus, which stimulates the release of luteinizing hormone by the pituitary gland. It is this hormone which promotes the final development and rupture of the follicles to release the eggs.

Consequently, the very first copulation may not lead to fertilization, and a male, not necessarily the original one, must copulate with her subsequently to ensure pregnancy.

SUPERFOETATION AND SUPERFECUNDATION

By mating females with different colour varieties of male mink it has been demonstrated that litters of mixed paternity can result. This is because ovulations continue after the initial fertilization, a phenomenon called superfecundation. As a result, even if a female becomes pregnant by a particular male she may continued to ovulate. Subsequent matings with a different male can result in kits being fathered by other males. If the female is mated more than once before ovulation, it is usually the sperm from the final mating that fertilizes most eggs (Venge 1971). Similarly, if a female has been fertilized by one or more males and 6–10 days later is mated by a different male, most of the kits born can be shown to have been fathered by the last male. This is because occasionally more eggs are fertilized than can implant and develop, and those most recently fertilized have the greatest chance of retention. This process is referred to as superfoetation (Shackleford 1952).

The female's physiology has evolved to maximize her chances of becoming pregnant. On mink farms the apparent level of male infertility runs at between 10 and 20% (Sundqvist & Gustafsson 1983). Although such a high level is unlikely in wild populations, females should nevertheless mate with more than one male to maximize the possibility of pregnancy. The mechanisms of superfoetation and superfecundation allow for this. The system requires a large number of ovulations, and induced ovulation helps minimize wastage of eggs. It also encourages the persistence of the males, and increases the opportunities for inter-male competition for paternity. The fertility of male ranch mink is highest at 2–3 years of age. The mating system of wild mink makes it difficult for sub-adult males to get mating opportunities. Since female fertility increases from the first to the second litter, males should invest more effort in mating and guarding adult females, and tend towards promiscuity, particularly at the end of the mating season.

The combined effect of these phenomena act to produce a very interesting situation regarding parentage of the offspring and the potential cuckoldry of the male who achieved the first mating. It is a situation in which we would expect the evolution of mate-guarding by the first male following his preliminary mating attempts. Only if he remains in attendance with the female and successfully defends her against other males for 48–72 h can he be sure of paternity of any kits. Because of superfoetation, that same male should return some days later to guard the female again. However, as we have seen, the male may have another female resident within his range and thus his dilemma becomes one of which 'mink-wife' to stay with, and for how long! If he abandons any female she is likely to be mated by a roving male who will probably achieve at least partial paternity. Increased mating effort at the end of the breeding season will be rewarding in terms of the number of kits sired. Male mink copulate for longer towards the end of the breeding season (Hansson 1947). Only after the female stops ovulating will the final paternity of her litter be decided.

Direct evidence of such mate-guarding by wild male mink has not been forthcoming to date. Examples of males consorting with females for short periods have been described by Ireland (1990) during studies on coast-living mink. The longest period of 'guarding' was for 2 days after mating—hardly sufficient to guarantee paternity. This is possibly the main reason why male mink do not show paternal care of the young. The risks of investing in another male's offspring are just too high. Only the female can be guaranteed that her genes are carried by the offspring and therefore she single-handedly rears the young.

DEVELOPMENT

After release of the egg from the ovary it is swept into the upper end of the oviduct. Fertilization takes place in this fine tube as sperm are wafted towards it by the action of cilia. The embryo grows from a single cell, dividing five to six times to form a hollow ball of cells (the morula) in about 6 days, after which it is released into the uterus.

DELAYED IMPLANTATION

Gestation in mink varies from 40 to 75 days, with an average of 50 days (see Fig. 8.2). Such variation is unusual in mammals generally, but is displayed by many species of mustelid, and other mammals including the seals.

Early theories attributed the variable pregnancy length to delayed

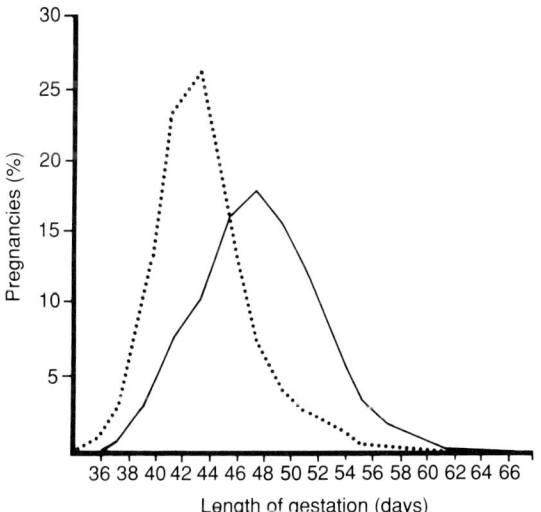

FIGURE 8.2 *Variation in the duration of gestation for females mated once (solid line), and for females remated 7 days later (dotted line). Source: Joergensen (1985).*

fertilization, or a highly variable period of foetal development. It is now well documented that the situation arises from a period of delayed implantation, during which foetal development is suspended at the morula stage. Instead of implanting when it reaches the uterus, the embryo remains detached and dormant for up to 30 days. Only after ovulation has ceased will the embryos implant. The result is synchronized foetal development, despite the protracted period over which fertilizations may have occurred. Throughout the population most births occur during the first week of May.

Changing day length seems to be the stimulus which eventually causes implantation, since gestation can be shortened by artificially increasing the day length or lengthened by decreasing the day length following mating. The balance between the hormone progesterone, secreted by the ruptured follicle (corpus luteum) in the ovary, and oestrogen, produced by the follicle before rupture, is important in determining when implantation occurs. It will not occur until the corpus luteum becomes active in April and begins to secrete progesterone.

The timing of birth is critical, since the young must be born during a season of relative food abundance if they are to have a good chance of survival. Additionally the provision of food for the young can put a severe strain on the female since the cost of pregnancy and lactation are high and the food resources within her range do not expand to cope with these additional hungry mouths. Delayed implantation thus provides a mechanism whereby the date of mating and fertilization can become independent of the time of birth, and also allows for the period of superfecundation. Hence, the rut can occur at a time when the males are in prime body condition, and the females can still give birth at a time of food abundance.

Implantation involves the embryo attaching to the wall of the uterus. Once the embryo is attached, the placenta forms and the embryo completes its development in 28–31 days.

PARTURITION

The birth season is usually from the last week in April to mid May. On mink farms in the northern hemisphere there is a peak in births during the first week in May. This also seems to be the case for wild populations in the UK.

Hormones from the pituitary, ovaries and possibly the placenta are involved in initiating parturition. Again the interplay between the levels of progesterone and oestrogen are responsible. Oestrogen was dominant during ovulation, while progesterone was effective in promoting implantation and the maintenance of pregnancy. Now the two hormones act as antagonists in their control of uterine contractions. Oestrogen has a stimulatory effect and progesterone an inhibitory effect on muscular contractions. The increase in oestrogen levels at this time prepares the uterus for the expulsion of the foetuses. The duration of parturition is unknown since most births seem to occur at night. The placenta is also expelled, and this is eaten by the female resulting in black tarry droppings 8–12 h after birth.

LITTER SIZE

Litters as large as 17 have been reported for ranched mink, but generally litters of 10 or more are infrequent (Enders 1952). It is believed that the disposition for litter size may be heritable. Enders reported one female to have produced 32 kits over 3 years. The more typical litter size ranges from four to six kits. Figure 8.3 shows that little size may well differ from young to old mothers. The percentage of barren females is smaller and the litter size tends to be larger in older females. The date of birth also seems to have some effect on the ultimate size of the litter with later litters generally being smaller.

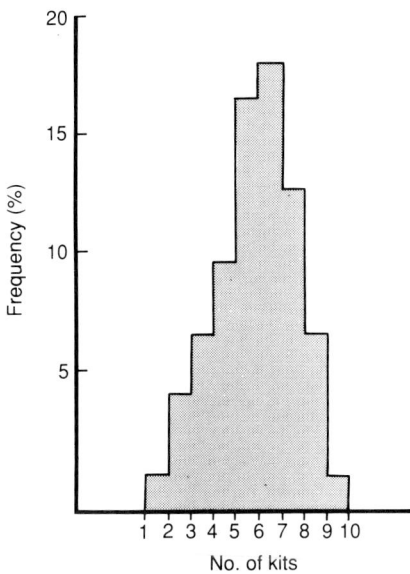

FIGURE 8.3 *Average litter size in ranch mink (old females). Source: Joergensen (1985).*

Various sources of information are available to give data on fecundity. Palpation of the abdomen of live-trapped pregnant females allows the number of foetuses to be assessed, the number of active nipples is an indicator of the number of kits being fed, and in post-mortem material, the presence of uterine scars tells us how many embryos were present before birth. Chanin (1983) estimated the fecundity of female mink in Devon to be between 5.6–5.8 kits, whereas in Sweden, Gerell (1971) estimated litter size as 3.6–3.8 kits based on direct observation. This may imply a high level of early kit mortality.

In our wild Scottish population of coast-living mink the number of kits produced was only known with any reasonable degree of certainty for three litters (Ireland 1990). One female was observed with three kits in 1983, and in 1984 she had four. In May, palpation of the abdomen of a pregnant female revealed the presence of at least four foetuses, but some time after parturition only two or three kits could be heard squeaking in her den, and by June only one kit was in evidence. Nipple counts provide a reasonable measure of the number of kits a female is feeding, and this can be supported by subsequently trapping the young at the time of dispersal. Table 8.1 suggests two to three kits per female as the typical level of production of wild mink even in this coastal area of food abundance and relative security from persecution.

TABLE 8.1 *Litter sizes of eight female mink recorded on the Ross Peninsula over 5 years.*

Year	Female number							
	6	7	10	11	18	24	25	27
1982	4/–	2/–						
1983	5/4		–/1	3/3				
1984				3/4	1/2			
1985				3/2		4/1		
1986						4/1	4/4	2/2

The first figure represents estimation from counts of palpable nipples on captured females and the second value represents actual captures of kits prior to dispersal.
Mean number of kits (nipples) 3.1 per female.
Mean number of kits (captures) 2.0 per female.
Source: Dunstone & Ireland (1989).

GROWTH OF THE KITS

Young mink are deaf, blind, bald, and helpless. From a birth weight of about 5 g they rapidly increase to about 100 g over the first month. The first vestiges of a coat appear after the first week. This lengthens and thickens to form the juvenile coat by the end of the first month. It is not until the first or second week in June (5–6 weeks old) that the kits' eyes open, and by this time they are highly mobile and may emerge from the den only to be dragged back to the safety of the nest by their mother.

The rapid growth of mink kits makes great demands on the mother's milk production. Apart from biochemical analyses of mink milk, there have been few investigations of lactation. The milk appears to have very high fat content. Up to 3–4 weeks the kits are entirely dependent on her milk for nourishment, after that time she will bring prey back to the den. On the ranch, mink kits are weaned at 6–7 weeks.

Comparisons between the sexes reveal that while the relative growth rate of male and female kits is similar for the first 1–2 months, thereafter marked sex differences develop. Males have a consistently higher weight gain than females (Fig. 8.4). On a ranch these sex-related differences in weight gain can appear as early as 21 days after birth. Growth rates of males and females between 11–17 weeks old were 10% and 7%, respectively (Gregory 1987). In

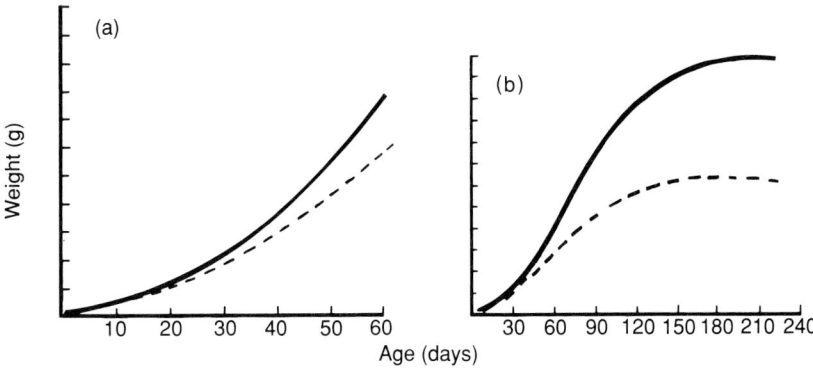

FIGURE 8.4 (a) Growth curve for the early development of ranch mink kits. (b) Growth curve for development of ranch mink. Solid line, males; dotted line, females. Source: Joergensen (1985).

feeding trial experiments conducted using natural foodstuffs (minced rabbit), this difference was not apparent until slightly later, and growth rates for males and females were only 3.2% and 2.6% respectively.

REPRODUCTION OF THE EUROPEAN MINK

Female European mink also become sexually mature at 9–10 months old, and in the male mature sperm are present from February to April. The breeding season however, seems to be considerably later than for the American mink. Oestrus, which lasts between 3 and 6 days, begins between 20 March and 1 June. Most of the effective matings occurred at the end of April. The oestrus state of the female is more easily recognized than in American mink since the vulva swells and can be pinkish-lilac in colour. Apparently there is no delayed implantation and gestation lasts from 40 to 43 days. Litter size ranges from two to seven kits. Development of the kits is similar to American mink, although the teeth appear earlier. Ternovsky (1977) notes the litter to break up at 2.5–4 months.

DISPERSAL

In Scotland the first time a kit was observed outside the den alone was at approximately 6–7 weeks old. On Ross Island, Ireland (1990) observed kits making short trips to the shore with their mother in July, and on one occasion he watched at she dragged a herring gull carcass to where they were sheltered under a rock, perhaps indicating that they were being weaned at this time. By 26 July the kits were out alone or in pairs on the shore, swimming and making short duration dives of up to 5 s. The kits were in the process of being rejected by the female at this time. One was ejected violently when it visited its mother's resting place, after which the mother moved to a different part of the island. He further observed the kits to become independent of the mother in August when none were denning with the mother.

CHAPTER 9

Population Biology

AGE AND SEX RATIOS

Analysis of the population structure reveals the ultimate product of the various reproductive and social strategies. For the ecologist, the determination of the age and sex ratio will enable an understanding of the success of reproduction and juvenile survival. It may suggest to what extent competition favours experience or youthful vigour, and will allow the relative importance of the various causes of death to be assessed. For the management agencies, it is also crucial to understand the population age structure if a successful harvest of fur is to be gained and 'over-exploitation' avoided. Similar criteria are rather differently applied to the management of pest species to achieve effective control, since here 'over-exploitation' is the aim. Thus, it is because the mink can be either a valuable fur-bearer or a pest species (depending on the population) that considerable attention has focused on the methods of accurately determining the age of individuals within the population.

TECHNIQUES FOR DETERMINING AGE

Various techniques for determining the age of mink have been developed; these are of varying accuracy and use in field-based studies. Some only allow relative age to be determined. Often information concerning the proportion of juveniles in the population is all that is desired. On other occasions it may be important to know the precise age structure of the population so that a life-table (Table 9.1) can be compiled. This may then allow times and causes of mortality to be pin-pointed.

Only subjective assessment of age can be made on the evidence acquired from living mink in the wild. Often it is only possible to classify them to 'juvenile', 'adult' and perhaps 'very old'. The most accurate method of determining the population age structure is to capture and mark mink as kits.

TABLE 9.1 *Mink population age structure as determined from a sample of specimens[a] collected on Vancouver Island, British Columbia, between 1968 and 1973.*

Age[b]	Cumulative proportion (%) of animals		
	Males (n = 58)	Females (n = 37)	All (n = 95)
0–1	35	54	43
0–2	66	82	72
0–3	86	91	88
0–4	98	97	98
0–5	100	100	100

[a]Animals collected, trapped, killed by automobiles, or dying in livetraps throughout the study period.
[b]Number of adhesion lines in the periostium of dentary bone.
Source: Hatler (1976).

POST-MORTEM DETERMINATION OF AGE

It is possible to accurately age mink using a variety of laboratory techniques, however none of these are suitable for use in the field with live animals since they involve post-mortem examination of skeletal and dental material.

Growth rings in sectioned teeth and jaws

Teeth are composed of a layer of enamel covering cementum and dentine which surrounds a central pulp cavity. Cementum is continually formed around pre-existing tooth material. Environmental effects, usually seasonal, result in fluctuations in the deposition of these substances which cause incremental lines to develop correlating with regular events in the animal's life. Consequently, the number of these lines or 'growth rings' corresponds to an estimate of the age of the animal. In the relatively stable environment of the mink farm these growth rings may be less apparent.

In carnivores the canine and carnassial teeth best show the phenomenon. These are extracted, or taken with a section of the jawbone (see Fig. 9.1). They are then chemically decalcified using weak acid (e.g. formic acid), stained (e.g. haematoxylin), embedded in wax and thin-sectioned in the conventional way on a laboratory microtome (see Morris 1972). Alternatively, ground sections can be prepared direct from the hard tooth. Slices are cut with a saw and these are rubbed down to thin sections on an oil-stone.

The sections are usually taken in the longitudinal plane of the tooth. Microscopic examination reveals growth rings as light and dark tissue which originally resulted from seasonal differences in feeding and hormonal activity. The clarity of the bands depends upon the degree to which the animal was affected by the prevailing environmental conditions. Figure 9.2 shows the location of rings in the tooth and mandibular bone of ranch mink. It has been suggested that the light-staining material (cementum) is deposited from

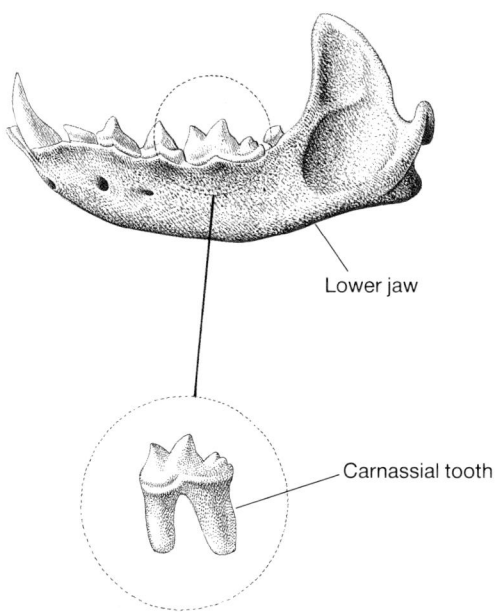

FIGURE 9.1 *Region of the lower jaw and carnassial tooth used in age determination.*

November to January, while the incremental line develops from March to April.

Pascal & Delattre (1981) tested this method on a sample of 27 farm-raised mink and found that for older animals, at least, it was a good indicator of their true age. Growth lines in the jaw bone have also been used. However, Chanin (1983) found them less satisfactory when attempting to verify the ages of wild mink caught on a Devon river, since juveniles occasionally possessed equivalent numbers of growth lines as adults.

Bacular morphology
An alternative way of distinguishing adult from juvenile male mink is to examine the penis bone (os penis or baculum). This method has also been extensively used as a relative ageing technique for other mustelid species.

The conformation and size of the baculum provides a reliable way of distinguishing adult and juvenile mink (Elder 1951, Lechleitner 1954). The changes are brought about by the increased flow of testosterone at sexual maturity. The bacula of adult mink (Fig. 8.1) are more massive with a heavier shaft and greater development of the proximal end. Those from young animals lack the characteristic ridge formed at the point of attachment of the corpus cavernosum in the bacula of adult males. These criteria of conformation, weight and length appear to hold true for distinguishing adult

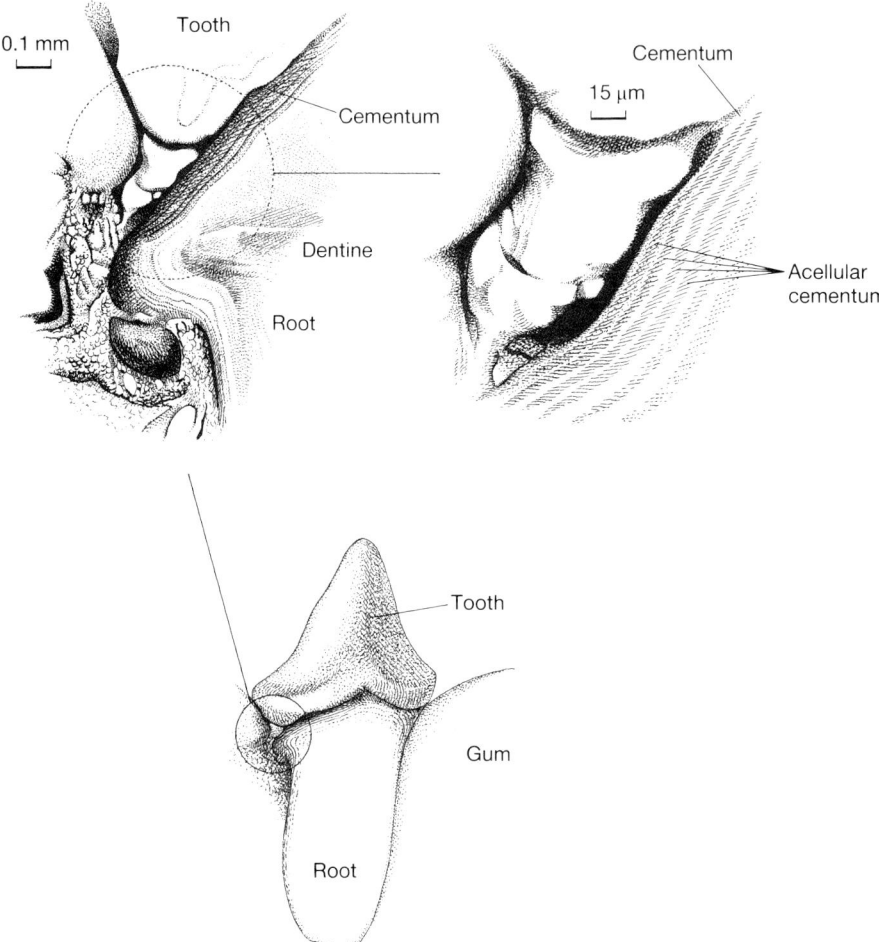

FIGURE 9.2 *Rings in the cementum of a carnassial tooth. Source: from Pascal & Delattre (1981).*

from juvenile males in both ranch bred and wild mink (see Fig. 9.3), although there can be some overlap. Observations of the bacula of feral Irish mink (Fairley 1980) suggested that the bone could continue to grow through to at least March of their first year. Elder (1951) considered the weight of the baculum to provide a more accurate indication of relative age than its length.

Skull sutures
Skull measurements are too variable to permit accurate classification of adult and juvenile animals partly because of the overlap between males and females. Birney & Fleharty (1968) have demonstrated that distinct, compared to absent or feint, sutures between the nasal and premaxilla and the

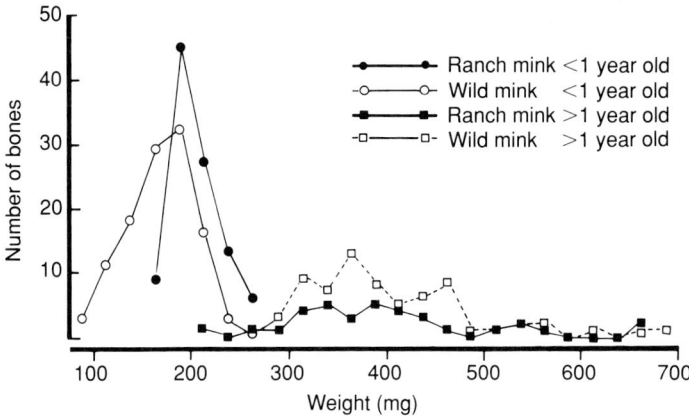

FIGURE 9.3 *Frequency distribution of baculum weight from 170 ranch mink of known age, and 144 wild mink. Source: Lechleitner (1954).*

jugal and zygomatic processes of the squamosal bone can be used to distinguish juvenile from adult mink, respectively. Chanin (1983) has used this character to classify adult male and female mink successfully.

Eye lens weight

The dry weight of the eye lens has been used to categorize spotted skunks into three age classes. An early attempt to use this criterion on mink (Birney & Fleharty 1968) proved unsuccessful, probably due to poor or inappropriate preparation techniques. More recently, Pascal & Delattre (1981) have shown that the lens weight increases linearly with age up until 1.5 years. The technique is of limited usefulness thereafter.

SEX RATIOS

In general, mammals have a sex ratio of 1 : 1 at the time of conception. A ratio significantly different from this in a population at large is indicative of ecological factors affecting survival or longevity of one of the sexes relative to the other. Sex ratio is normally estimated by extrapolation from a sample of the population. This technique is thus dependent on an unbiased sampling technique, that is, one that is equally likely to capture males or females.

Trapping studies of mustelids commonly indicate sex ratios that differ significantly from 1 : 1. Usually the sample is biased towards males. However, this most probably arises because of the greater home range size and increased mobility of males, at least during the mating period. Males range further and travel more rapidly than females, and are therefore more likely to encounter traps (Buskirk & Lindstedt 1989). Trappability may vary from month to month, leading to pronounced variation in the apparent sex ratio (Table 9.2). In a long-term study, Chanin (1983) captured 16 males: 15 females and 23 males: 16 females respectively, on two rivers in Devon, while in Scotland (Dunstone & Birks 1983, Ireland 1990) trapped 49 males and 33 females over a 6-year study period. With these reservations on the empirical findings, it is more likely that the sex ratio is equal.

TABLE 9.2 *Trapping effort and trapping success, as total number of mink caught and the total number of different individuals captured.*

		Captures			
		Total		Individuals	
Month	Trap nights	Male	Female	Male	Female
January	294	35	6	11	5
February	448	38	17	16	9
March	550	86	15	31	8
April	439	20	3	13	2
May	247	7	0	4	0
June	306	3	0	2	0
July	390	13	13	8	6
August	546	21	33	14	22
September	356	16	11	7	11
October	171	14	9	8	7
November	324	18	12	9	8
December	341	31	10	10	7
Total	4412	302	129	49	33

Source: Ireland (1990).

The sex ratio as determined from fur-trapping seems to show less male bias than that determined from live-trapping studies. The sex of pelted mink can be determined from their skins by the presence of a penis scar in the male or nipples in the female. On this basis, Petrides (1950) observed a 1:1 ratio in a sample of 249 pelts from central Ohio. A similar ratio was obtained by Whitman (1981) from a sample of 147 mink also trapped in central Ohio.

Marshall (1936) proposed the analysis of tracks in snow as a means of censusing the mink population. Given their sexual dimorphism of body size males and females might be distinguishable on the basis of the size of their tracks. However, the intersexual overlap of body size, the presence of immature animals and the difficulty of accurately measuring tracks in snow limit the usefulness of the technique.

CENSUSING AND ESTIMATING POPULATIONS

The number of mink in a population can be estimated using mark–recapture techniques. This involves live-trapping mink in a given area, marking them with ear-tags or freeze-branding, and releasing them at the site of capture. The trapping programme is then repeated at a later date. Reasonably accurate estimates of population size can be obtained, as long as some inherent assumptions are not violated. One of the major assumptions made is that the population is a closed one, that is, that there has been no immigration or emigration, or births or deaths during the study period. Given the high mobility of mink, particularly males, at certain times of the year, population estimates based on these methods are not very reliable. Other methods have also been used, but they are even less reliable. These are mainly based on the abundance of field sightings, tracks and scats.

In areas where fur trapping is extensively carried out, trends in the status of populations can be obtained by analysis of fur harvest records. However, it is debatable whether this gives a true indication of the population density. For territorial species like the mink, the removal of animals from an area has the effect of drawing in those from adjacent ranges who take advantage of the vacant space. The larger the area studied the better (provided it is evenly sampled), as migration at the perimeter becomes less significant with increased area.

Fur harvests vary widely from year to year, not only in response to the numbers of animals, but also as a result of inconsistent trapping pressure and highly variable trapping conditions arising from climate. Thus, fluctuations in the fur harvest may not indicate variation in the numbers of mink on a year to year basis, but over a longer period can be used as a rough guide to whether the population trend is increasing or decreasing.

Typically it has been noted that there is a higher proportion of juveniles to adults (4.5:1) in a population subject to intensive trapping compared to an unharvested population (0.3:1, Mitchell, 1958). This ratio would be minimized in an unharvested population at or near the carrying capacity of the habitat, where the success of a juvenile is limited primarily by its ability to secure a territory.

The potential for population growth is estimated from the fecundity of female mink. As might be expected this has been the subject of intensive study on commercial mink farms. However, little is known of the fecundity of wild mink. Evidently the number of kits successfully weaned and dispersed is considerably fewer than her biological capacity as determined on mink farms. Many carcasses are available as a result of trapping operations, either for pest control or for the fur harvest. From these it should be possible to examine the reproductive tracts to determine the number of placental scars and hence the last litter sizes. However, since the fur harvest occurs outside of the breeding season, only limited information will be available from this source. The number of animals available from pest control operations has two annual peaks: in February to March (the rut), when the catch is predominantly of males, and in August to September (the time of dispersal) when the catch is predominantly of juveniles. The most accurate method of determining fecundity would be to trap whole litters, or to excavate breeding dens before kits disperse, but this is not generally possible. Hence, we have to resort to data from ranched animals and here successive litter sizes of individual mink have been found not to vary significantly from the first to the seventh year of breeding.

POPULATION DENSITY AND POPULATION DYNAMICS

Population density will vary according to several factors, for example, the carrying capacity of the habitat, intraspecific aggression and territoriality, and level of predation. Other variables affect the accuracy of the population estimate depending on the method used. These include variations in trapping pressure, trapper effectiveness and even weather conditions.

The highest estimates of population density come from areas with abundant and stable aquatic habitats, swamps and marshes. Estimates can be widely variable from year to year. In Montana, Mitchell (1961) obtained a mink density of one per 11.8 ha (280 mink in a 33 km^2 area), but the following year the catch was only of 109 mink (one per 30.3 ha). Other estimates range from one native American mink per 18.8 ha in Wisconsin to one per 625–909 ha for feral American mink in Sweden.

Differences in the population density of mink between two areas can result from various factors including availability of prey and competition with other predators, suitable den sites and social instability through persecution.

MORTALITY

We rarely know what disorders are associated with mortality in natural populations. Carcasses are difficult to find; perhaps dying animals retire to their dens, or get washed away by rivers or tide. Hatler (1976) determined the age of a sample of 95 mink which died from various causes (see Table 9.1). There was a preponderance of animals of up to two years of age.

PRE- AND NEONATAL MORTALITY

Three hypotheses have been advanced (Venge 1971) to explain the high levels of kit mortality observed in ranch mink. It is thought the same processes will be working in wild mink populations. In the early phase of reproduction two processes may reduce the viability of eggs. A long interval between mating and the release of ova may increase the number of unfertilized eggs. Failure of fertilized eggs to implant in the uterus may result from hormone imbalance. One of the major factors is undoubtedly the requirement for a second mating to ensure fertilization. This causes a new wave of egg release from the ovary. Only 15% of the original fertilized eggs are thought to survive this competition between embryos of different ages.

On fur farms there is a high number of still-born kits, perhaps amounting to some 4%. This may be followed by a heavy loss of kits within a few days of

TABLE 9.3 *Incidence of external pathological symptoms and ectoparasites in wild mink on Vancouver Island, British Columbia, 1968–1973.*

| | | | | Frequency of symptoms (nearest whole %) | | | |
| | | | | | Ectoparasites | | |
Class	n	Tooth symptoms[a]	Body problems[b]	Wounds[c]	Ticks	Lice	Fleas
Adult males	160	49	29	19	14	3	1
Juvenile males	99	61	10	17	14	5	2
All males	259	54	22	19	14	4	1
Adult females	70	46	19	39	16	4	1
Juvenile females	40	50	15	15	25	8	5
All females	110	47	17	30	19	6	3
Adults (both sexes)	230	48	26	25	14	4	1
Juveniles (both sexes)	139	58	12	17	17	6	3
All animals	369	52	21	22	15	4	2

[a] Animals with no apparent wounds or parasites.
[b] Having broken or excessively worn teeth (root canal exposed).
[c] Having open or healed wounds, broken bones or abnormal pelage.
Source: Hatler (1976).

birth, perhaps 7% on the first day post-partum, and representing a total loss of 15–16% in the first 11 days following birth.

PATHOLOGY

Of 369 mink examined by Hatler (1976) (see Table 9.3), 21% had some problem with their teeth, either through excessive wear or fracture. Another 22% showed evidence of body wounds, most of which had probably been acquired during interactions with other mink.

ECTOPARASITES (TICKS, FLEAS AND MITES)

Mink commonly carry light infestations of external parasites: ticks, mites and fleas. The species which have been found on native mink in America and feral mink in the British Isles are given in Table 9.4.

Of the fleas recovered, *Ctenophthalmus nobilis vulgaris* is a parasite of the mink's prey (mice, voles and rats) and *Nosopsyllus fasciatus* is a parasite of brown rats. Most tick infestations involved one or two engorged individuals embedded in or around an ear, on the head, or between the shoulder blades, but occasionally as many as six were found. *Ixodes hexagonus* is also commonly found on the hedgehog (*Erinaceus europaeus*), and other British mustelids, while *Ixodes ricinus* is the sheep tick, Hatler (1976) noted a high incidence of tail wounds in coast living mink (see Fig. 9.4). Some of these wounds were very large and many seemed to have large numbers of a small mite associated with them. The wounds seem to cause the mink considerable

TABLE 9.4 *Ectoparasites commonly found on native American and feral mink.*

Parasite	Location	Reference
Fleas		
Ctenophthalmus nobilis vulgaris	Ireland	Fairley (1980)
Nosopsyllus fasciatus	Ireland	Fairley (1980)
Typhloceras poppei	Ireland	Fairley (1980)
Paleopsylla minor	England	Chanin (1983)
Malareus penicilliger	England	Chanin (1983)
Megabothris walkeri	England	Chanin (1983)
Orchoppeas sp.	British Columbia	Hatler (1976)
Ticks		
Ixodes hexogonus	Ireland	Fairley (1980)
Ixodes ricinus	England	Birks (1986)
Ixodes spinitalpis	British Columbia	Hatler (1976)
Ixodes pacificus	British Columbia	Hatler (1976)

FIGURE 9.4 *Tail wounds. Photo: Dave Hatler.*

irritation since they often licked the afflicted area. Immersion in sea-water when hunting may have delayed the healing process. Hatler (1976) also found occasional infestations of biting lice (*Trichdectes* sp.).

ENDOPARASITES (TAPEWORMS, FLUKES AND ROUNDWORMS)

Only a small number of European studies have undertaken internal pathological examination of dead mink—none have reported the occurrence of internal parasitic worms, although undoubtedly they occur. In North

America Hatler (1976) has recorded a high incidence of such parasites in a population of coast-living mink. Over half of a sample of 75 autopsied mink suffered from infection with the sinus worm (*Skrjabingylus nasicola*) (Fig. 9.5). Some 15% of infected animals had sustained severe skull damage from the burrowing activities of the worm, probably enough to seriously affect the mink's behaviour. Infection rates were higher in juveniles and females. The infection has also been reported in Swedish mink.

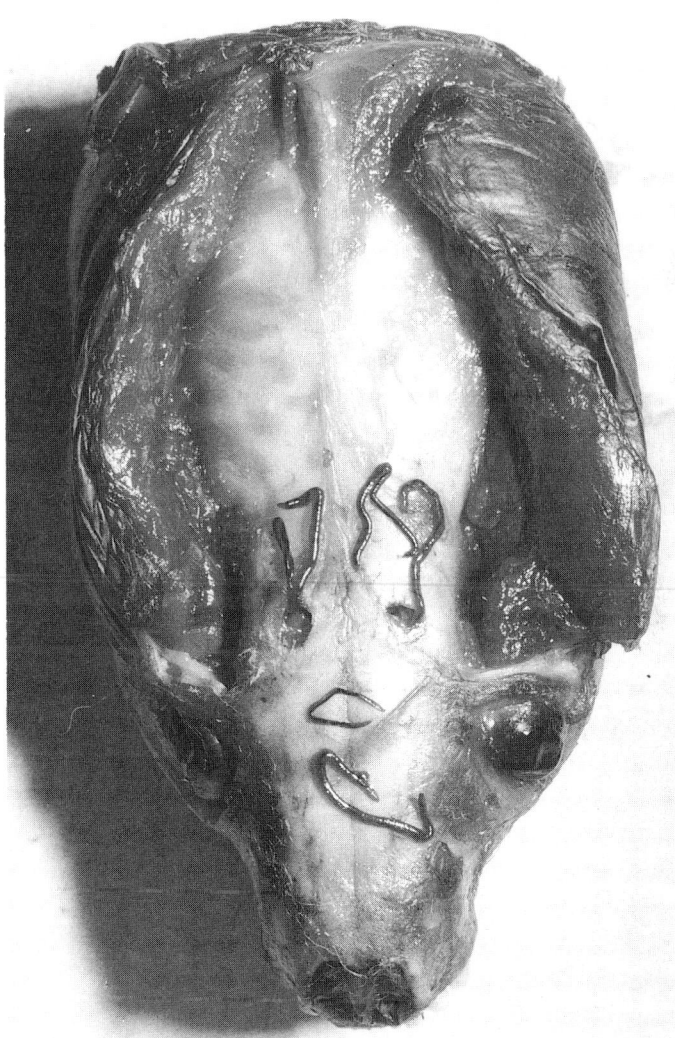

FIGURE 9.5 *Mink skull showing infection with the sinus worm* (Skrjabingylus nasicola). *Photo: Dave Hatler.*

The ascarid (Nematoda) worm *Baylisascaris devosti* was found in the intestinal tract of about 10% of the animals. Infections were generally light, but one animal had 163 worms, many more than 73 mm long, present in the gut. Infection with the acanthocephalan (Cestoda) was also quite light, although 50% of the animals were infected. Autopsy of 94 mink carcasses obtained from trappers in Ontario revealed approximately 25% to be infected with the kidney worm *Dioctophyma renale* (Cestoda). In some cases the infected kidney was grossly hypertrophied, weighing up to 50 g instead of the normal 3 g, and containing up to seven worms (Wren *et al.* 1986).

NATURAL PREDATORS

It has frequently been stated that the mink has no natural predators apart from man. This may the case in some countries where there are feral populations, but even here they are likely to receive some attention from raptors. In their native lands they are preyed on by wolves, foxes, coyotes and bobcats, hawks, owls and eagles. In Russia, wolves (*Canis lupus*), lynx (*Lynx lynx*), wolverine (*Gulo gulo*) and even otters are their principal predators. Eagle owls may be a significant avian predator. It is doubtful whether any of these predators have a significant effect on prey populations.

EFFECTS OF DDT AND PCBs

As a top carnivore in the food chain, mink may be liable to concentrate pollutants initially taken up by its prey. Ranched mink are known to be susceptible to mercury contamination of their fish food (Aulerich *et al.* 1974, Woebeser *et al.* 1976). Poly-chlorinated biphenyls (PCBs) are also known to be toxic at very low concentrations (Platanow & Karsted 1973, O'Shea *et al.* 1981). Mink are most likely to encounter these compounds during predation of fish from freshwater habitats. The short lifespan of wild mink, compared with otters, may also reduce the significance of PCB toxicity. More effective control of the release of these compounds into the environment recently has reduced the threat to mink and other wildlife in many habitats. However, these compounds are very resilient to breakdown and are present in the environment for a very long time.

EFFECTS OF ACIDIFICATION OF RIVERS

In inland parts of the southern-most counties of Norway, freshwater fish populations are strongly affected by acidification of the streams and rivers, and in many areas are now virtually extinct. This has lead to the decline of the mink population, not only because of the reduction in available prey, but also because acidification has lead to the liberation of other toxic metals (aluminium and heavy metals) which then accumulate in fish, and may become concentrated in their predators, the mink and otter (Bevanger & Ålbu 1986b).

CHAPTER 10

The Fur Trade

People have always hunted animals for their fur, to clothe themselves or to use as a commodity to trade. Nowadays there are many more practical substitutes for fur, and it has become a pure luxury. To maintain demand, its aesthetic appeal and symbolic extravagance is backed by a massive promotional campaign. In 1987, the sale of furs was worth some estimated $US2 billion dollars world-wide. Figure 10.1 shows the annual world-wide production of mink pelts from 1976 to 1986.

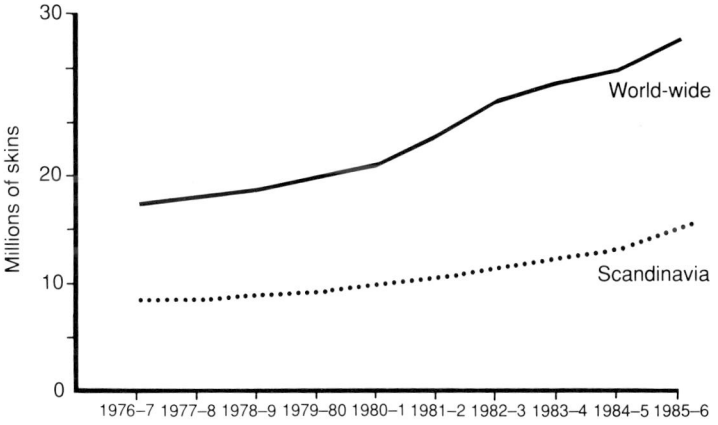

FIGURE 10.1 *Annual production of mink pelts from fur farms. Source: Lynx (1988).*

THE EXPLOITATION OF WILD POPULATIONS OF MINK

The exploitation of fur-bearing mammals was the incentive for much of the early exploration of North America. Trappers and traders penetrated all parts of the continent, and there were frequent conflicts between representatives of the fur companies from various European nations attempting to gain control of territories with good populations of fur-bearers and key trading routes. As long ago as 1670, the Governor and Company of Adventurers of England Trading in Hudson's Bay, later incorporated into the Hudson's Bay Company, began a trade in fur which they continued until very recently. Fur-trading is thus the oldest business institution in North America.

The number of animals killed world-wide for their skins is appalling. The total, excluding seals, was estimated at over 237 million in 1977–1978; a country by country breakdown is given in Table 10.1.

TABLE 10.1 *World summary of animals killed for fur (1977–1978).*

Wild animals	38 743 415
Captive-raised animals	26 999 550
Domestic animals	237 325 000
Breakdown of wild animals killed for fur by country (1977–1978)	
USA	18 784 261[a]
Canada	3 193 103[b]
Latin America	3 550 000
Europe	1 478 700
USSR (exports)	1 709 075
Asia	4 391 626
Australia	2 516 861
New Zealand	2 636 209

[a]1983–1984 figures for animals trapped in the USA, although incomplete, show 11 million trapped despite higher trapping pressure to meet increased demand. This is an indication of a 'limiting out' or local extirpation of fur-bearing species to meet demand for skins. Figures from International Association of Fish and Wildlife Agencies.
[b]Canadian 1983–1984 figure is 4 394 361 (WSPA).
Source: Nilson (1980).

Wild mustelids constituted a substantial proportion of the furs traded from North America in the 18th and 19th centuries. Wild-caught pelts of mink and otter were important to the economy of the northern lands. Early explorers traded directly with native Indians for furs, but later white settlers and trappers, who came to America to make quick fortunes, became the major supplier of animal pelts. However, the initial rich plunder became a lean living as the animals were over-exploited. Often trappers had to move to even more remote places when the wildlife was depleted locally. As we have already seen, for the sea mink (*Mustela macrodon*), once common along the coasts of Maine, New Brunswick and Nova Scotia, over-exploitation probably led to extinction. Populations of many fur-bearers, particularly the larger and slower breeding mustelids were greatly reduced by the early 1900s. The reduction in numbers of pelts of these highly prized, economically important species stimulated much ecological research on population dynamics. This eventually lead to the imposition of closed seasons. All Canadian provinces and 47 American States allow at least a limited harvest of mink. Although few States conduct population estimates, the lack of a consistent downward trend in the number of wild mink pelts harvested (Fig. 10.2) is taken to indicate that the populations are not over-exploited.

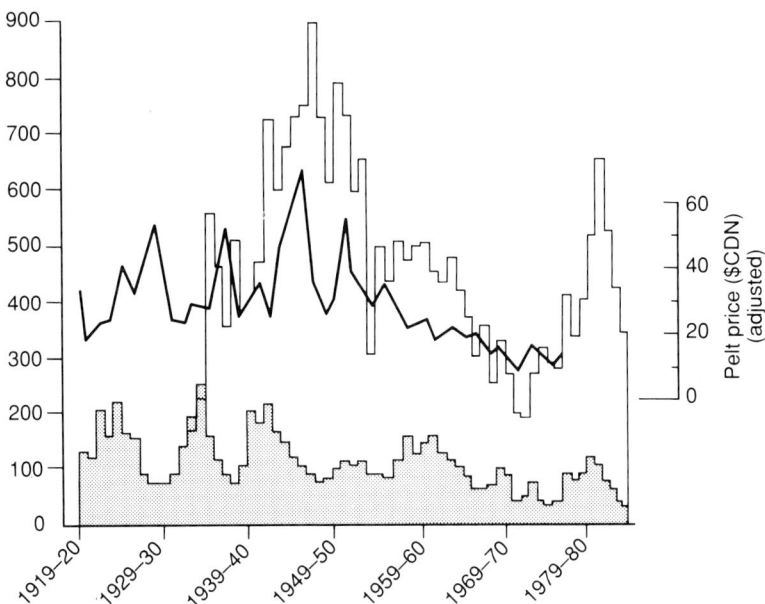

FIGURE 10.2 *Sales of mink pelts in Canada (hatched bars) and the USA. Solid line shows adjusted average price in Canadian dollars. Source: Eagle & Whitman (1984).*

The USA traps more animals for their fur than any other country, with the possible exception of USSR. In 1977–1978 some 18.8 million individuals of 23 species were captured in the USA. In the present day only a small proportion of this trade is of wild mink. US fur harvest data indicate some 300 000–400 000 mink are trapped annually with the bulk coming from Minnesota, Wisconsin, Alaska and Iowa. In Canada, analysis of statistics for 1985–1986 show approximately 88 000 wild mink pelts with a value of $Can 3 million, in addition to nearly 1.5 million ranched animals being marketed. The distribution and harvest density of wild mink in Canada and the USA for the 1983–1984 trapping season is shown in Fig. 10.3.

Most wild mink currently trapped in the USSR are from released populations of the American mink. It is likely that the demise of the European mink is directly attributable to over-exploitation for its fur.

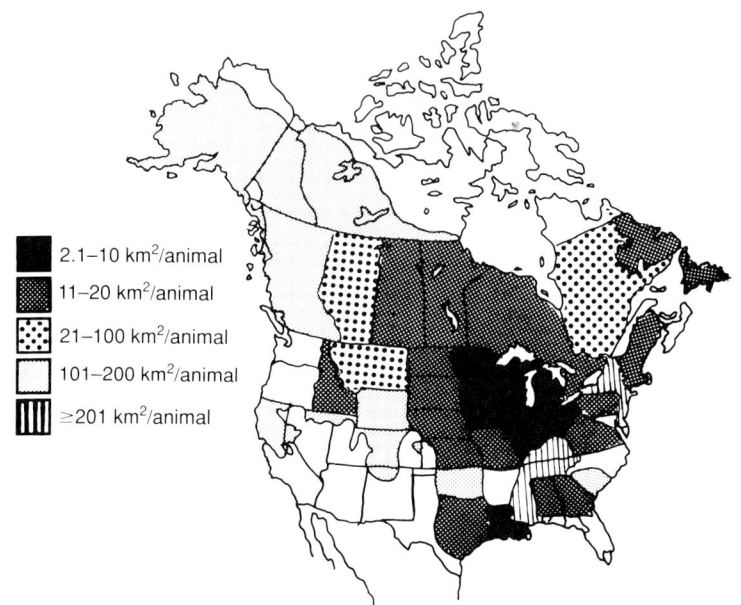

FIGURE 10.3 *Distribution and harvest density of mink in Canada and the USA for the 1983–1984 trapping and hunting season. Source: Eagle & Whitman (1984).*

Following World War I, there was a boom in the fur industry in the USA, when the wearing of animal furs became a fashion rather than a necessity. High pelt prices led to the severe depletion of wild mink, marten (*Martes martes*), fisher (*Martes pennanti*), river otter and beaver (*Castor canadensis*) populations. This led to the acceptance of ranch-bred fur.

*O*RIGIN OF RANCH POPULATIONS

Mink raising is a relatively new industry compared with other branches of farming. The earliest attempts to raise mink in captivity date from the early 1900s in Canada. Initially the fur trade would not contemplate using pelts from captive-bred animals as the skins of wild-caught mink had denser underfur and better resistance to fading. This led to the experimental breeding of mink to promote better fur quality.

From the outset, little attention was paid to the origin of the mink. The foundation breeding stock was usually trapped in the neighbourhood of the farm. Of the 11 subspecies of American mink, the three main contributors to the genetic stock are *Mustela vison vison*, *M.v. melampeplus* and *M.v. ingens*. Captured animals were fed, housed and kept in good health until they produced young. Three types of ranching have been attempted: extensive range, whereby the animals are kept under conditions as close as possible to nature within large enclosures with free access to water; the colony method, whereby many mink are kept together in large houses; and thirdly, the pen system, widely used today, where each mink is housed individually within the confines of a small cage and nest box.

THE BREEDING OF 'MUTATION' MINK

Despite its initial disadvantages, the ranching of mink quickly developed into a prosperous industry. There are several reasons for this success, including the large litter size that can be reared in captivity, the short time to maturity and the relative ease with which colour varieties may be produced and maintained in the breeding pen.

Native American mink are dark brown in colour. The original intention was directed towards the production of fur equal in quality to that of wild mink. Selective breeding was imposed to promote darkness of the fur and fineness of texture. As mink were reared in increasing numbers, freak animals or mutations began to appear. These mutation mink arose by the in-breeding of small captive populations, and cross-breeding of mink subspecies from a number of different geographic areas. The Silverblu or Platinum was the first colour variety to be noted in captivity, and is the aberrant colour variety most frequently noted in feral populations in the UK. Conceivably this colour mutation may have also arisen in wild populations of mink in the USA, but none was captured alive for breeding purposes. The first recorded appearance of a Platinum mink occurred on a Wisconsin farm in 1931 and led to the establishment of the Whittingham strain of Silverblu Platinum mink. This was the first commercially important mutation mink, and the consumers' enthusiasm for it was largely responsible for the intense competition to develop new colour varieties.

The colour of the fur arises from pigment granules within each shaft of guard hair. These granules vary in size, number, shape, arrangement and in the intensity of pigment they contain. These slight differences (particularly in number and size of the granules) are wholly responsible for the plethora of colour varieties that form the basis of mutation mink breeding.

The science of heredity (genetics) is the study of the transmission of characteristics from parent to offspring. In reality it is not the characteristics themselves that are inherited, but the genetic disposition to react in a certain way to the environment. For the mink farmer to predict the offspring he is likely to obtain as a result of crossing different types, it is necessary to know the ancestry of the parents, their genetic constitution and at least a simplified knowledge of the principles of Mendelian inheritance.

THE PEA AND THE PRINCESS'S ROBES

Gregor Mendel, a 19th century Austrian monk, formulated a series of laws of heredity which determined the outcome of crossing different varieties of pea. Although they were soon to be used in plant breeding, their practical importance in the selective breeding of domestic animals was not realised until 1930. Their value in the planned production of new colour types of mink was soon realized and gained extra impetus from the high prices paid for these novel pelts. Much has been written on the practice of breeding mutation mink. Our knowledge of the science underlying this often lucrative husbandry is due to Shackleford (1957).

The basis of Mendelian genetics is that each individual acquires a complete set of genetic information from each of its parents. It thus has two copies of every gene: one maternal and one paternal. Each gene may exist in several forms in the population: the different forms are referred to as alleles. The two copies received by the offspring may be of the same or different alleles. When it, in turn, passes to its offspring, it may transmit either one or the other: for each unit of genetic material the choice is made separately, so that the final combination of genes transferred is a matter of chance. The character displayed by the individual depends on the dominant form of the allele. The recessive form will only be expressed when it is present in both copies. Sometimes alleles are 'co-dominant', producing a character intermediate between the two. Since Mendel's time it has become known that the genetic information is carried in the 'chromosomes' which are visible under the microscope when the cells are in the process of dividing to form gametes or body cells. The number, size and shape of the chromosomes is characteristic for each species of animal (or plant), and is referred to as the 'karyotype'. The mink has 30 chromosomes (Fig. 10.4), 15 of paternal origin and a matching (homologous) set of 15 from the mother.

When the sex cells—the sperm or the ovum—are formed, each homologous pair is separated into two different cells. The number of chromosomes in this basic set contained in each sex cell is called the 'haploid' number, denoted

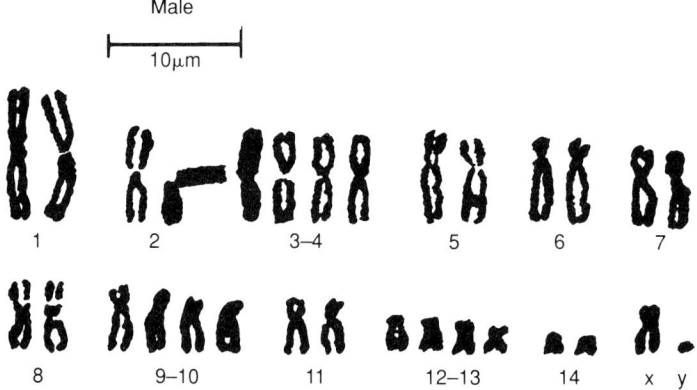

FIGURE 10.4 *The mink's chromosomes ordered in pairs by size, the sex chromosomes are indicated* x *and* y. *Source: Venge (1971).*

'n'. When two sex cells combine, every cell in the resulting animal will contain the full, 'diploid' set, that is '2n'. Thus, for mink, 2n = 30. The understanding of this process gives a physical basis for Mendel's theories, and also explains why some genes (namely those on the same chromosome) and the characters they control are usually inherited together. This linking of genes in chromosomes may seem to reduce the amount of potential variation, but it still leaves 2^{15} (32 768), possible combinations in each sex cell, and when two such sex cells combine there are 4^{14} (1084 million) possible chromosome combinations. Genes may also be recombined, or swapped, between homologous pairs of chromosomes, so over several generations even the linkage patterns may be scrambled. Little wonder that even litter mates can show considerable variation.

BREEDING FOR COAT COLOUR

The coat colour of the mink results from the interaction of at least 18 genes. When each gene is present in its normal *wild type* form, the mink will display its natural standard dark coloration. But if any one of the genes is altered, causing dysfunction or reduced expression over all or part of the body, a deviant coloration may be produced. Such mutations occur naturally at very low frequencies, and often go unnoticed because the action of the normal gene inherited from one parent masks the alternative form inherited from the other parent. In this case, the mutant gene is said to be 'recessive', and its effect will only be seen in an animal that inherits the same mutant form of the gene from both parents. (Different forms of the same gene are known as 'alleles'). Individuals that inherit the same allele from both parents are said to be 'homozygous' for the character controlled by that gene; individuals with two different alleles are referred to as 'heterozygous'. This is why deviants are more likely to appear in very 'in-bred' populations, where the parents are related to one another.

Occasionally mutations occur that cause a changed coat colour even when there is only one copy of the mutant gene present. This is called a 'dominant' mutation. Similarly, alleles that give an intermediate effect in the 'heterozygote', compared with the pure-bred (homozygote) for each, are said to be 'co-dominant'.

Five of the currently known mutant genes are dominant; of the recessive genes, five produce grey colours, six produce brown, four pale brown to tan, and four produce white mink. All but one of these recessive mutant genes produce solid coat colours. In contrast, all but one of the dominant mutant genes produce a patterned effect on the pelage. As a rule, the effects of a combination of two mutant genes is to produce a coat colour in which the effect of both parents can be distinguished.

It is obvious that you can't always tell an animal's genetic make-up by just looking at it. For this reason, a geneticist makes a clear distinction between the apparent type of animal (its 'phenotype') and its genetic type (or 'genotype'). Thus, a mink which has the normal dark brown coloration may be classified as phenotypic *wild type*, although its genotype may possess a recessive allele for a different coat colour. The breeder can only discover this genotype by crossing the animal with another that possesses the recessive gene, and observing the coat colours of the offspring. From the Mendelian model of inheritance, the breeder can predict the ratios of different colours he would expect from a cross between any two genotypes. But the colour of each kit is determined separately by chance, so it is not surprising that the actual ratio of kits within one litter is often very different from the prediction. Only when the frequencies are averaged over a large number of similar crosses will the Mendelian ratio prevail.

A number of examples of the standard crosses are explained in Figs. 10.5–10.8. By following through the possible combinations of alleles, it is easy to see how the predicted ratios are derived. The genotype of each animal is denoted by pairs of letters, with each gene of interest assigned a different letter. Thus, in the first example, the *Silverblu* gene is denoted p. A capital

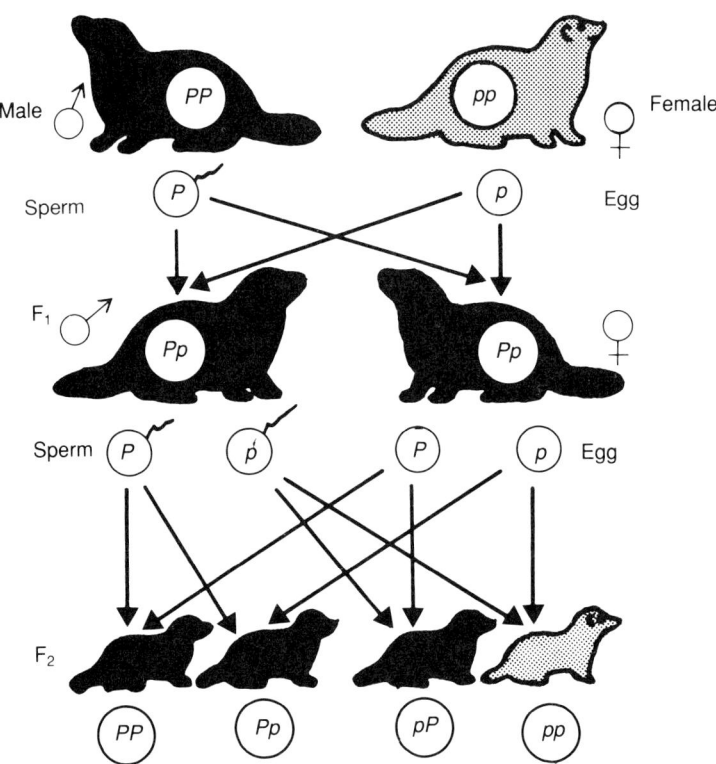

FIGURE 10.5 *Diagrammatic representation of a cross between a homozygous* Standard *(PP) and a* Silverblu *(pp), and the offspring which have been mated with each other to produce the F_2 generation. Source: Shackleford (1950).*

letter P denotes the dominant allele (usually the *wild type*), while the lower case p indicates the recessive allele conferring *Silverblu* coloration.

Recessive mutation

The first example is that of the popular *Silverblu* mink, which is a classic recessive mutation. If a *Silverblu* animal is crossed with a pure-bred (homozygous) *wild type*, all the offspring of the first generation (the so-called 'F_1') will have the heterozygous genotype Pp. Note that all will have the normal dark brown phenotype, but will be carriers of the *Silverblu* gene. If two such carriers are crossed, the alleles can recombine in a number of ways. The expected genotypic ratio is 25% pp, 50% Pp, 25% PP, that is a 1:2:1 ratio. But as the Pp and PP individuals look the same, the phenotypic ratio will be 3:1 (*wild type*: *Silverblu*). If the heterozygote is crossed to a pure *Silverblu*, an equal number of *Silverblu* (pp) and *wild type* carriers (Pp) will result. Such a cross of a hybrid with one of its parents is called a 'back-cross' (see Fig. 10.6) and is a quicker way of determining the carrier status of a brown coloured animal.

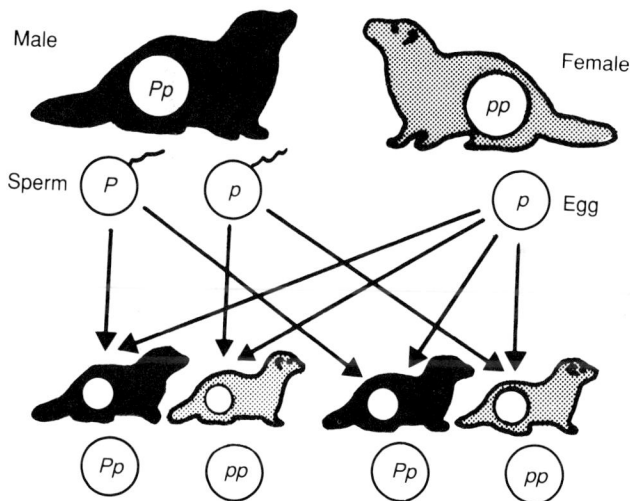

FIGURE 10.6 *Diagrammatic representation of a back-cross between a heterozygous standard (Pp) with a Silverblu (pp); 50% of the progeny will be of the Standard type and 50% Silverblu. Source: Shackleford (1950).*

Recessive mutations have the advantage that a *Silverblu* mink will always breed true. Once a pure-bred population has been established there is no need to introduce other types. But a breeding programme like that described here can be used to widen the genetic base to avoid inbreeding, or to quickly establish a pure-bred stock from a few animals of a valuable colour variety.

Co-dominant mutation

Sometimes one gene may be only partially dominant over the other. A simple example of this is seen in *Black-cross* mink, where the coat is white with a line of black fur running from head to tail and with a transverse line running across the shoulders. The *Black-cross* gene is denoted S. The pattern is manifested in the heterozygous condition Ss. The homozygous condition SS is almost completely white. If two of the heterozygous animals are mated the F_2 generation is 25% white SS, 50% *Black-cross* Ss, 25% black ss offspring (see Fig. 10.7). The breeder of such a heterozygote colour must keep pure-breeding stock of both parental types. However, such a gene has the advantage that there are no hidden carriers: you can always tell the mink's genotype from its colour.

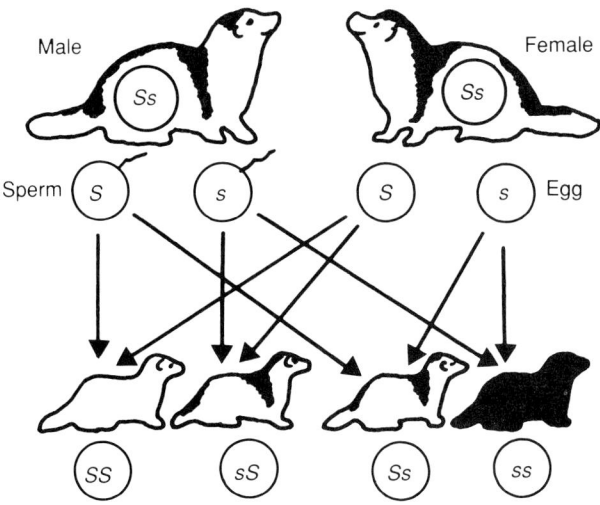

FIGURE 10.7 *Diagrammatic representation of Mendelian segregation where two* Black-cross *mink (Ss) are mated with each other. The result is one homozygous* Black-cross *(SS), two* Black-cross *(Ss), and one completely pigmented (ss) offsping. Source: Shackleford (1950).*

Complementary genes and dihybrid crosses

So far we have only considered colours governed by a single gene introduced into the normal genetic background. Sometimes two separate genes act together in an additive or complementary fashion, to produce a new colour. An example of this is the *Sapphire* mink, a pale blue-grey, produced by combining the *Silverblu* (pp) gene with the slightly darker grey *Aleutian* gene (aa) into the same animal. Note that the *Silverblu* mink will possess the normal, dominant version of the *Aleutian* gene (AA), while the *Aleutian* possesses the normal allele (PP), homologous to the *Silverblu*. The cross is thus (ppAA) × (PPaa). All of the offspring are 'dihybrids' (AaPp). The two genes would seem to cancel each other out, as all the kits have the standard brown colour. In fact they are mutually hidden, and a cross between the two dihybrids will segregate the two alleles to give nine possible genotypes, and the four colours: *wild type*, *Silverblu*, *Aleutian* and *Sapphire* (see Fig. 10.8). The prized *Sapphire* mink will be a rare one in sixteen kits (aapp). However, once a few pure-breds are obtained, the stock can be quickly built up by back-crossing.

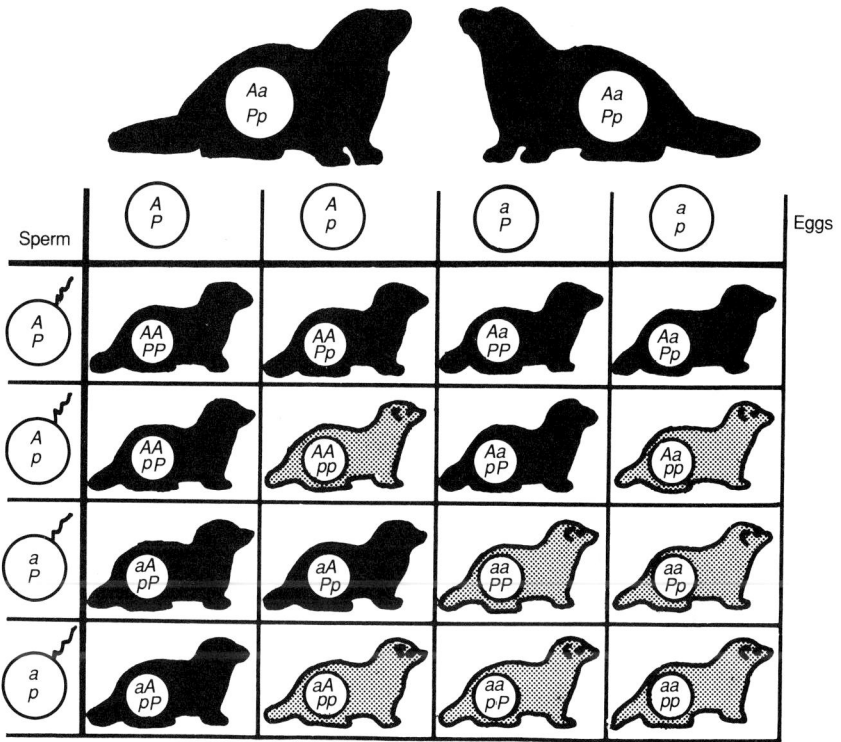

FIGURE 10.8 *Dihybrid segregation. Heterozygote carriers (AaPp) × (AaPp) are mated giving 16 possible combinations of offspring (see text). Source: Shackleford (1950).*

By the same method it is possible to use tri-hybrid crosses to produce, for example *Violet* coated mink of genetic constitution aappmm, and even quadruple recessive types, such as *Moylambersapphire* (marketed as *Blush*, actually pink in colour) mink of genetic constitution mmrraapp.

Genes may interact with one another in some instances to suppress or enhance a characteristic. In other cases, for example, *Blue-frost* gene, the homozygotic condition is unknown since it results in the death of the individual carrying it at an early stage of foetal development. Single genes may confer more than one effect to the phenotype, an phenomenon referred to as 'pleiotropy'. The *Aleutian* gene used in the production of *Sapphire* also confers in the homozygous condition (aa) reduced resistance to aleutian disease (not the disease itself), small body size and low fertility. The deafness of *Hedlund* white mink is also a pleiotropic effect. This condition makes mating difficult and the females are poor mothers, presumably because the mink are incapable of responding to the vocalizations of their offspring. In animals that are homozygous for the *Stewart* gene, all surviving male offspring are sterile and the females of low fertility.

During the crossing of mink to produce these established colour varieties of

TABLE 10.2 *The principal gene symbols.*

Wild mink-standard	AA	BB	CC	ee	HH	KK	MM	nn	PP	RR	ss	TT	ww
Mutants													
1. Aleutian	aa												
2. Royal pastel		bb											
3. Albino			cc										
4. Finnblack				Ee									
5. Hedlund					hh								
6. American palomino						kk							
7. Moyl							mm						
8. Jet black								Nn					
9. Silverblu									pp				
10. Ambergold										rr			
11. Wild glow										$r^d r^d$			
12. Blackcross											Ss		
13. Shadow											$S^H s$		
14. Socklot												$t^s t^s$	
15. Svensk palomino												$t^P t^P$	
16. Finnwhite												$t^w t^w$	
17. Nordisk buff												$t^n t^n$	
18. Stewart													Ww
Combinations													
19. Sapphire	aa								pp				
20. Moylsapphire (violet)	aa						mm		pp				
21. Ampalosapphire	aa					kk			pp				
22. Ambersapphire (hope)	aa								pp	rr			
23. Pastel cross		bb									Ss		
24. Palosapphire	aa								pp			$t^P t^P$	
25. Albinopastel (regal)		bb	cc										
26. Socklotpastel		bb										$t^s t^s$	
27. Socklotpastelsilver		bb							pp			$t^s t^s$	
28. Ampalosilver (pearl)						kk			pp				
29. Shadowsilver									pp		$S^H s$		
30. Shadowpearl						kk			pp		$S^H s$		
31. Jetsilver								Nn	pp				
32. Jetpastel		bb						Nn					

Source: Joergensen (1985).

mink, a farm may be fortunate enough to note a new mutation for coat colour. In the hey-day of the fur trade when the breeding of new colour varieties of mink was 'the be all and end all' of mink farming, the rancher would attempt to built up his stock of the new type as quickly as possible. Perhaps one or two kits in a litter would bear the desired characteristic in the homozygous condition. Working on the knowledge that the parents must be carriers (heterozygous) and that some of the siblings probably carry the same genes the farmer instigates a series of crosses to build up his stock. This might involve mating a male of the new mutant variety to its mother, sisters and even half sisters to produce more individuals of its type. Additionally, it can be mated with unrelated females to produce more carriers which can be subsequently back-crossed. Since the male can easily mate with five females during a breeding season the stock can be built up quite quickly. The process is slower, however, if the mutant individual is female, since, although she can be mated to her father or brothers, she will only produce one litter in a season. Inevitably the process involves the mating of close relatives and this can give rise to unfavourable, deleterious gene combinations affecting the health and vigour of the offspring. Such degeneration is universally associated with inbreeding.

The genetic constitution controlling coat colour in a variety of ranch mink varieties is shown in Table 10.2. A typical distribution of colour varieties of mink bred on farms in the UK (data from 1984) is shown in Figure 10.9. Most of the mink kept in farms around the world are 'brown' in colour, for example, *Dark/Jet, wild type* or *Pastel*.

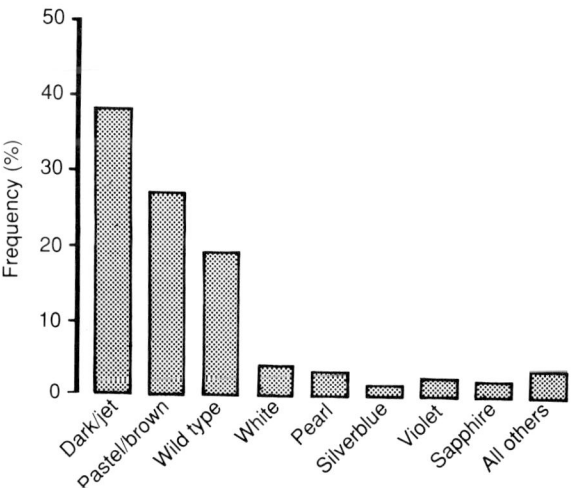

FIGURE 10.9 *The distribution of colour varieties of mink bred in the UK. Compiled from a census of 75 000 breeding females during March 1984. Source: Dunstone (1986).*

RANCH MANAGEMENT

The production of fur on a commercial scale is a high-risk business, not only in terms of fluctuating markets and the whims of the fashion trade, but also because of fluctuations in the availability of foodstuffs.

NUTRITION OF RANCH MINK

The food requirements of carnivores make them more expensive to keep than other animals domesticated for man's use. All commercially-bred fur bearers have a high protein and energy demand and the mink is reputed to be highest of all. To reduce the high cost of protein inputs, feeds have been developed which, in composition, are often far removed from the natural diet. Initially, 'wet-feed' mixes were used, comprising poultry offal, slaughterhouse by-products and fish waste (trimmings from filleting factories), even ground-up mink carcasses together with cereal and milk products. In Britain and parts of Europe this is still the practice on many ranches. The past 20 years has seen the development of artificial dry, pelleted foods especially in the USA and Scandinavia. These consist of a cereal base with fishmeal or livermeal or dried milk as the protein component, to which has been added fats and oils to raise the energy content of the ration. In the UK the waste oil from fish and chip shops is still occasionally used, and even waste broken biscuits may be incorporated to provide the cereal component!

In the UK, local availability usually determines the diet of ranched mink, but typically the diet includes a mixture of fish waste and slaughterhouse products, such as sheep's paunches and chicken waste products, to which are added cereal, milk products and energy-rich oils and fat (see Table 10.3). On mink farms in the Far East mink are reputedly fed on whale and seal meat. It is now commonplace on American and Scandinavian farms to feed a dry pelleted diet compounded from fishmeal and cereal (see Table 10.3).

THE RANCH MINK'S YEAR

This is illustrated in Figure 10.10. The most critical month of the year for the mink farmer is March, as this is the mating season and will determine the success of his operation for the entire year. To maximize breeding success in the farm environment, a mating programme has been worked out, largely by trial and error, which attempts to overcome the husbandry difficulties imposed by the reproductive physiology of the mink. Mating usually commences about 5–10 March on UK farms. The farmer determines who will mate with whom, and for how many times. As in the wild, mating is a violent affair. If the female is not on heat, a situation difficult to recognize because of the lack of visual signs, she will fight off all advances of her chosen suitor and the two have to be separated. Deaths can occur if the animals are left together.

TABLE 10.3 *Ingredients of diets for ranch-bred mink.*

Ingredients	Percentage
100% dry diet	
Protein	
Fishmeal	28
Livermeal	3
Skim-milk powder	4
Yeast	10
Soya bean oil meal	15
Carbohydrate	
Cereal grains	25
Fat	
Lard	15
Typical wet-feed used in the UK	
Fish waste/whole fish	35–40
Poultry offal	35–40
Slaughterhouse products	15–20
Cereal	10–20
Milk products	2–3
Fat/oil/vitamins	

Source: Dunstone (1986).

She is then re-tried every 2 days until a mating is achieved. Each male may be paired with up to five females over the season.

The female ovulates some 48 h after copulation. Her periods of heat last for only 48 h and will occur again 8 days later. To achieve fertilization it is essential to obtain a mating during a heat period so the female is re-mated, usually but not necessarily, to the same male, 7 clear days after the first copulation, and again the next day. This regime holds true until 18 March (in the UK), after which she is tried every day until a mating is achieved and then re-mated the following day.

During April, preparations are made for the birth of the kits, most of which are born in the first 2 weeks of May, just as in feral and native populations. The kits are weaned in June and separated in August. September sees the start of the 'furring-up' process (moult) prior to pelting in November to December when the fur is prime, that is, has reached its maximum length and density, but has not yet started to lose its colour. The coats of wild mink remain prime for 8–9 weeks.

At this stage the farmer has the maximum number of animals present on the farm, the young in brother–sister pairs, and the adults each in a separate pen, all being fed the maximum quantity of the best quality feedstuffs. Most mink are just 8 months old when killed and pelted. The number of animals maintained on farms is highly variable, ranging from less than 100 breeding females to over 5000 (see Fig. 10.11). The latter can easily result in a stock of 20 000 animals being housed at pelting time.

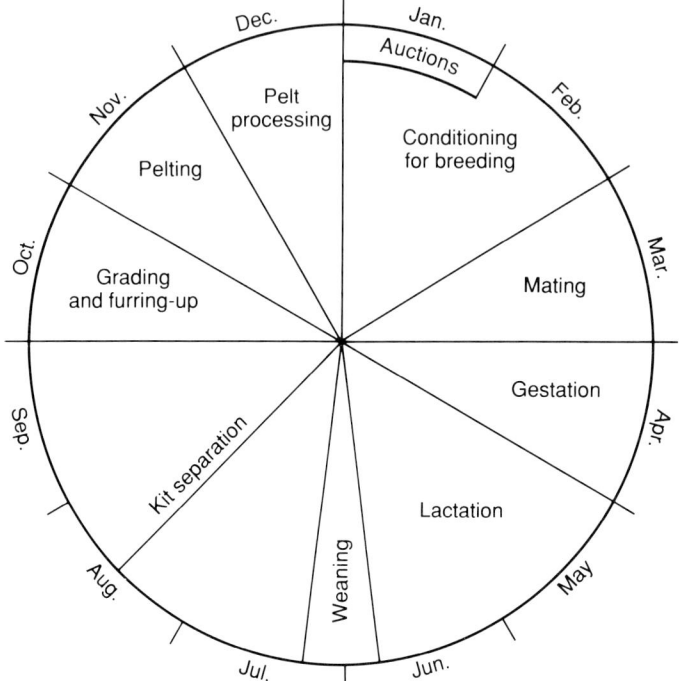

FIGURE 10.10   *Significant events in the mink farmer's year. Source: Dunstone (1986).*

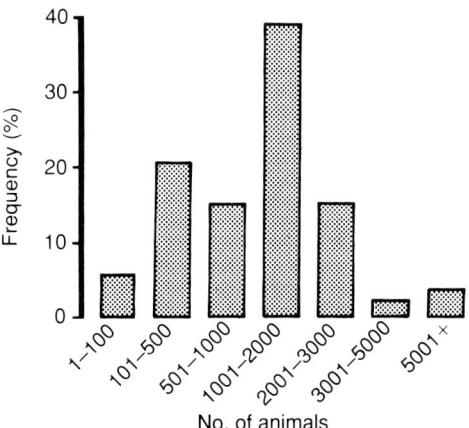

FIGURE 10.11   *The size distribution of mink farms in the UK. Compiled from a sample of 54 farms during March 1984. Source: Dunstone (1986).*

FUR PROCESSING

Mink are small animals and considerable numbers (40–60 pelts) are required to make one fur coat. Mink are killed, commonly by a lethal injection of barbiturate or chloral-hydrate, or by gassing with carbon monoxide fumes. The corpses are skinned cased, that is inverted like a glove, without being opened down the belly. The pelt is then fleshed to remove most of the fat and lanoline-rich grease from the skin, a process aided by rolling it in sawdust for 30 minutes. Finally the pelt is air-dried on a board at between 18 and 20°C for 48–72 h, after which the complete batch is sent for auction or some alternative means of marketing.

PRODUCTION AND MARKETING

Mink pelts have a value at all times of the year, but reach their greatest value when prime. Slight differences in quality (colour, density, nap) can produce wide differences in value. No two pelts are identical, but are matched as far as possible into bundles, a bundle is sufficient to make a garment. The true value of these is established when a number of buyers appraise the bundle and bid for it in a free competitive market. The individual mink rancher has a number of alternative ways of disposing of his crop, but most will eventually pass through the hands of the auctioneer. Until recently the Hudson's Bay Company of London was the main auction house for European production. Following its closure, the Danish Fur Sales of Copenhagen has taken its place. Russian pelts are auctioned in Leningrad and Leipzig.

The rancher and buyer both benefit from the auction system. The former because his pelts are presented to many potential buyers, and the furrier because he can inspect and compare many thousands of well-matched pelts displayed under ideal conditions. The final advantage of the auction system is the inter-sorting of pelts from many different ranches into the same bundle, thus allowing even the small rancher to sell his skins at their true marketable value.

From the point the pelts leave the farm there is little the individual rancher can do to influence the price he will receive. Marketing organizations can enhance the final price by advertising campaigns and by promoting orderly marketing procedures. In North America two such organizations exist: the Great Lakes Mink Association (GLAMA), founded in 1941, and the Mutation Mink Breeders Association (EMBA). These were launched to promote each mutation type to the fur-trade. In Scandinavia, SAGA furs provides a similar function. These cooperatives undertake advertising campaigns, and a policy of management of sales in tune with what the market can absorb. In the UK, the London Fur Group performs a similar function and the Fur Breeders Association organizes an annual conference and other educational events.

In the USA most furriers are located in New York, but the business is in decline, contracting from 2500 companies in the 1940s to some 400 in the late 1980s. This does not necessarily signify a reduction in the market, rather, pelts are being shipped around the world to take advantage of cheap manufacturing costs and then re-imported to Europe for sale.

Mink farming is a huge world-wide production industry. Well over 23 million pelts pass through western European auction rooms each year (see Table 10.4). Relatively few animals are now reared in the UK; in 1981 the total production was 245 000 pelts from 67 farms. In 1992 there are just 27 farms still operative. Similarly, only a small proportion of coats are sold here. In a reversal of the traditional trade, North America is now the main consumer of European pelts. Russia produces in excess of 10 million pelts annually, the bulk of which are marketed and used domestically.

TABLE 10.4 *World production of mink in 1987–1988.*

Country	Million
Denmark	10.20
Finland	3.80
Sweden	2.00
Norway	0.51
Holland	1.60
France	0.55
Spain	0.25
Italy	0.38
Austria	0.05
West Germany	0.35
Belgium	0.17
Great Britain	0.26
Ireland	0.10
USSR (export only)	4.10
East Germany (export only)	0.32
Poland	0.08
Czechoslovakia	0.05
China (export only)	3.10
USA	4.70
Canada	1.50
Japan	0.68
Korea	0.05
Total	34.80

Source: Mink Factories *publ. Lynx and CIWF (1988)*.

THE WELFARE OF RANCH MINK

Assessing the welfare status of any species is difficult and often involves subjective judgement of what an animal requires. This itself is often clouded by what we would feel if we were confined under similar conditions. The application of such sentiments to other animals is referred to as anthropomorphism. The fur industry would argue that production, measured by the level of breeding success, is an adequate indicator of well-being, whereas anti-fur groups might consider mortality and/or the incidence of abnormal behaviour as a more appropriate estimator of welfare status. If welfare is evaluated on the basis of longevity, breeding success and reduction of stress, then there is little to commend life in the wild, since wild mink live shorter lives, raise smaller litters and experience stress from periodic shortage in resources, for example, food, mates and the aggression-fraught business of acquiring and controlling a territory. If, however, you assess welfare on the quality of life experienced, then the wild animal's free access to the basic necessities of life, the possibility of satisfying its natural inquisitiveness and instincts to swim, hunt, procure a territory, and pass on its genes to future generations, must surely be an improvement over months and in some cases years of captivity.

Few guidelines have been published on the welfare of ranched mink, and opinions vary considerably on housing requirements and management regimes. None appear to be based on sound behavioural or physiological studies, or take any account of the ecological data concerning wild mink such as that presented in this book, even though much of this information has been published for some considerable time. However, care must be taken in extrapolating standards from the wild animal to its caged counterpart. Who is to judge if the welfare of the wild animal is good?

Few attempts have been made to intensively breed and rear carnivores. Only mink, ferrets (fitch), sable and silver foxes have been successful. One of the main arguments against fur-farming is that the mink is, at best, incompletely domesticated and hence its welfare requirements are considerably different than those of familiar domestic animals. Mink differ from such animals in two fundamental ways. Firstly, they are solitary animals, whereas most domestic stock are derived from gregarious species and remain so. Secondly, they have been in captivity for a comparatively short time, some 60–80 generations since capture from the wild. The nature of the selective breeding that has been carried out also differs. The strong selective pressure can leave little scope for the concurrent selection of those animals which behave most suitably in their new restricted environment. There seems to have been little attempt to breed for docility or lack of aggression towards conspecifics—common features of most domesticated species.

The Farm Animal Welfare Council (an advisory group to the UK government) has listed five basic needs for all livestock kept in captivity:

(i) freedom from thirst, hunger and malnutrition,
(ii) appropriate comfort and shelter,
(iii) the prevention of, or rapid diagnosis and treatment of injury, disease and infection,
(iv) freedom from fear,
(v) freedom to display most normal patterns of behaviour.

To a certain extent, the animal's best interests are also those of the farmers. We should hope that conditions (i) and (iii) are catered for if the farmer is not to lose his animals. In most instances ranch mink are protected from environmental hazards including climatic extremes, are fed and watered and given treatment for ailments. The remaining requirements are more difficult to assess.

CONFINEMENT

Virtually all of the mink's activity in the wild is dictated by its requirements for survival and reproduction. Labours of necessity they may be, but the dilemma of the welfare debate is how much of this activity and stimulation is also required to maintain a sense of well-being, to avoid boredom and disorientation. If physical requirements are catered for on a farm, does the animal then show a reduced need for activity? It is unlikely that there is a biological requirement to swim, but caged mink show considerable levels of locomotor activity, frequently including abnormal behaviours such as pacing and circling. Some time ago we conducted some preliminary experiments involving supplementing the diet of a free-ranging, but radio-collared mink by providing rabbit carcasses for his consumption. This provision of an excess of food had little effect on his nightly wanderings even though he no longer needed to leave his den to forage.

In one respect at least the ranch mink resembles the wild animal. Both show a low level of overall activity, perhaps 2–4 h per day. The rest of the time is spent asleep or resting. The nature of the activity shown is very different and the reasons for the inactivity of the two types of animals is probably different, energy conservation on the one hand compared to boredom on the other.

The largest cages encountered on mink farms rarely exceed 30 wide × 45 high × 90 cm long, and many are considerably smaller. Even this size of cage is likely to be used to house more than one individual. It is indeed remarkable that an animal that, in its native land or in feral populations, roams over kilometres of riparian, lacustrine or coastal habitat will successfully breed under such conditions of confinement. The fact that escapee animals so readily adapt to an alien countryside, and to differing species of prey and climatic conditions indicates that man's attempted domestication has had an almost negligible effect on this species.

Perhaps ironically, the saving grace of fur-breeding is that most of the animals do not have to endure this confinement for long, since the majority are killed and pelted at the age of 7–8 months.

ISOLATION

Given the solitary lifestyle adopted by wild mink outside of the breeding and kit-rearing seasons, it has been argued that the presence of conspecifics in close proximity will be stressful. Certainly all ranched mink appear less stressed if they are given a nest box into which they may retire. Separation of females by visually opaque screens reduces weight loss during pregnancy, indicating that the sight of another individual may be stress-inducing. It has also been noted that adjacent animals adjust their activity patterns so as not to be simultaneously active. Various experiments have been carried out to investigate the effect of visual isolation on breeding success, and these seemed to indicate that females so kept, raised larger litters (Gilbert & Bailey 1969, 1970), although more recent experiments (Hoffmeyer and Møller 1986) have been somewhat contradictory. Perhaps the intervening 20 years of 'domestication' has had an effect. The separation of the kits from their mother should be to a different part of the farm rather than to adjacent cages, thus reducing stress in this instance. Given the mink's dependence on olfactory stimulation in detecting and even recognizing conspecifics, the overwhelming aroma (!) prevalent on a mink farm could also induce stress. The same is probably true for audition. Certainly smell and hearing predominate in the mink's sensory world.

BEHAVIOURAL DISORDERS

The incidence of behavioural disorders is suggestive of a welfare problem. The Commission of the European Community (CEC report 1983) details four types of abnormal behaviour:

(i)   injurious behaviours,
(ii) stereotyped behaviours,
(iii) abnormal body movements,
(iv) apathetic behaviour.

Caged mink can exhibit all of these (Jonge *et al.* 1986); needless to say wild mink (feral or native) exhibit none. Particularly evident are stereotypic running up and down the cage and circling behaviour which occurs just before feeding time. The occurrence of injurious behaviours, of which the most common are tail sucking and tail biting, can be frequent.

CONGENITAL DEFORMITIES

Post-mortem analyses of 1249 mink kits which were either stillborn or died within 1 day of birth revealed 89 (3.8%) to be deformed. Of these, 70% suffered from dropsy, 20% from hernias (cerebral or umbilical), 7.9% from cleft palate or hare lip and 1% from cyclopia. It is impossible to get comparable information from wild populations.

DISEASES

On mink farms it is extremely important to minimize the incidence of diseases and parasites, particularly since even if death does not occur, the animal's loss of condition may adversely affect the fur quality. Mink are healthy vigorous animals; they rarely get sick if they have adequate food, water and clean conditions. Because of their 'dirty' habits, and the manner of feeding, ranch mink are prone to botulism unless extreme care is taken with the origin and treatment of foodstuffs. Outbreaks of botulism can wipe out an entire herd. Nutritional diseases include steatitis, caused by high levels of unsaturated fatty acids and a deficiency of vitamin E. Wet belly (urinary incontinence) is probably caused by a mineral imbalance and can ruin the pelt. Chastek's paralysis is an interesting disease, caused by the presence of an enzyme in certain species of fish (carp), which when included in the foodstuff can result in the destruction of vitamin B (thiamine).

Blue colour varieties of mink, including Aleutian and Sapphire, are susceptible to a viral disease, Aleutian disease, so called because the original losses were only recognized in these varieties and the disease was thought to be hereditary. Now it is known that these colour varieties have a greater susceptibility to the disease. Females pass on the disease to the unborn kits in the uterus. This has been possibly the most economically important of the mink ailments. Nowadays ranchers test for the disease and by culling infected animals have largely bred it out of the stock. Mink are commonly innoculated against viral enteritis, botulism and distemper.

In the 1940s and 1950s fish from the Great Lakes was a prime source of protein for local mink farms. During the middle 1960s they found that they could no longer use this cheap abundant source because it resulted in poor production and high levels of kit mortality. Analyses of Coho salmon from Lake Michigan revealed 15 ppm of poly-chlorinated biphenols (PCBs) from pesticide production. Laboratory tests showed mink to be extremely sensitive to PCBs; 30 ppm was lethal to adults, 15 ppm caused reproductive failure, and even 10 ppm depressed the growth of kits.

MORTALITY IN CAPTIVITY

Mink can live a long time in captivity. I kept one animal, which was used in behavioural experiments involving swimming and diving, for 8 years. Most ranch mink are pelted at 8 months, but a few breeders may be retained for two or three breeding seasons. The 'natural' lifespan of farmed mink is not known. Kit mortality seems to be high, with perhaps 9–12% being born dead, or dying within the first 3 weeks of life. The mortality of the offspring of first-year breeding females was greater than for older breeders.

THE ETHICS OF WEARING FUR

The ethical argument against wearing fur is that it is an unnecessary luxury product, for which no animal should be trapped or farmed and killed.

> Some who reflect upon this subject for the first time will wonder how such cruelty can have been permitted to continue in these days of civilisation, and no doubt if men of education saw with their own eyes what takes place under their sanction, the system would have been put to an end long ago.
>
> *Essay on Fur* Darwin (1878)

Long ago has become long hence, and still the issue is unresolved, often clouded in prejudice and misunderstanding on both sides. The trapping of wild animals for their pelts is impossible to justify in terms of conservation or animal welfare. Over 80% of America's wild-caught fur-bearers are taken in steel-jawed leg-hold traps. For mink this represents some 300 000 annually in the USA alone. Very few trappers are 'professionals' deriving their livelihood from trapping. In the USA, some 86% of trappers are under 20 years old. Weekend trapping by townspeople is also common. Leg-hold traps suitable for mink are relatively cheap at around $US5 each, and can be easily obtained from stores or mail-order companies.

The use of the leg-hold trap has been banned in some 64 countries, including Britain, because it is cruel and indiscriminate, but only a handful of North American States have even restricted its use. For every mink coat (comprising some 65 skins), it is estimated that some 180 'trash' or 'non-target' animals will also have been killed. Animals caught in leg-hold traps may often seem to suffer only minor injuries; however, X-rays show them to suffer bone fractures, torn muscles and ripped tendons. Frequently they chew off their own limbs to escape. Atkeson (1956) reported 27.6% of 209 trapped mink 'wringing-off' or 'chewing-off' their own limbs or escaping with the trap still attached to their limbs. Most trap-held animals will damage their teeth in trying to escape from the trap. Broken teeth, gum and jaw damage are common in such animals. The use of rubber-lined leg-hold traps has been advocated by the fur-trade, but these seem to be infrequently used, and their benefits regarding the welfare of the trapped animal are probably negligible.

To many people the ranching of fur is unacceptable because the animal is raised for its pelt alone. Other domestic species, farmed under equally intensive conditions, do provide subsidiary compounds. Only occasionally are additional products made available from the mink farm. Lanolin rich mink oil and a variety of compounds used in perfumery are available. The mink carcasses may be used in the production of fertilizers, but are more commonly incinerated or tipped at land-fill sites. In any case, the argument that such an industry would be justifiable if it produced items of utility as well as luxury would seem rather thin. The industry would not exist but for the fur, and the nature of additional by-products is as irrelevant to the mink farmer as it is to the mink!

Another argument in defence of fur-farming is that the demand for natural fur is insatiable, and the humane ranching of mink would relieve the wild population of some trapping pressure. However, we must consider the extent to which supply affects demand, whereby promotion of farmed fur products ensures a buoyant market for trapped pelts also.

At the moment there is very little legislation applying to the manner in which mink are housed. Training guidebooks for fur farmers emphasize production rather than welfare. In the UK, the Fur Breeders Association has produced a Code of Conduct for their members. This has now been adopted by a number of European countries. Unfortunately, not all breeders are members of the Association and not all members seem to adhere to this voluntary code.

For many years there has been considerable debate over the ethics of wearing animal pelts. From a practical standpoint, synthetic furs may have some advantages over real fur garments. They are moth proof, water repellent and considerably less expensive to buy, yet almost as warm, although possibly not as long lasting and durable as the real article.

However, few mink coats are worn for their warmth and durability alone. On a world-wide scale the propaganda campaigns of the anti-fur lobby are beginning to bite. Despite this, the market for furs in some countries continues to increase in line with their growing affluence and 'westernized' ideals. As a result, global pelt production is rising. It is difficult to say when saturation point will be reached, as any drop in price due to over-production brings the product within the reach of new consumers who had never before considered buying them.

Public opinion is rapidly changing at least in the UK and some other countries in Europe. In a survey of 2000 people conducted on behalf of Lynx (a UK based anti-fur campaigning group), 71% agreed (49% strongly, 22% slightly) that 'it was wrong to kill animals for their fur to make clothing'. Over 70% agreed that there should be a complete ban on the trapping and the rearing or farming of animals for their fur.

CHAPTER 11

Interactions with Man and other Animals

> Our children of tomorrow could curse us if they can never see a rabbit, an otter, or any other species of British mammal or waterfowl in the British countryside, which could be if the mink menace is not eradicated now.
>
> *Express and Echo* (31 October 1979)

Such was the fearsome reputation commonly held for the immigrant mink in Europe, and particularly in the UK. Although now largely discredited, enmity towards the mink is deeply ingrained, and many people would still sympathize with the sentiment expressed above.

Feral populations of animals often cause severe ecological problems, but it is the economic damage resulting from their effect on man's activities that usually attracts public attention. Many of the mammal species in the UK which are considered to be economic pests are also introduced species that have become feral. The Destructive Imported Animals Act (1932) prohibits the unlicensed keeping of specific non-indigenous species and empowers the Ministry of Agriculture Fisheries and Food to destroy such animals when they are found at large.

Efforts at eradicating feral populations of mammals in Britain have been successful with only three species: the muskrat (*Ondatra zibethicus*) which established colonies in Perthshire, the Welsh Borders and Surrey/Sussex in the 1930s; Himalayan porcupines (*Hystrix brachyure*) in Devon in the 1970s; and coypu (*Myocastor coypus*) in East Anglia in the 1980s.

It is now some 60 years since American mink were first imported to Europe for farming on a commercial scale, and just over 30 years since they were first recorded breeding in the wild. Despite early doubts concerning the viability of feral populations and a Ministry of Agriculture scheme to eradicate them, there has been a dramatic colonization of a large proportion of the waterways in the UK.

The pest status of the mink has been the subject of vociferous scientific debate for nearly 10 years. The landowners' view was eloquently expounded by Sir Christopher Lever in an emotive, but anthropomorphic article published in the popular scientific press (Lever 1978a) which was largely based on broad generalizations, but was widely quoted and singularly effective in provoking antagonism to the mink. The mink's corner was defended by Linn & Chanin (1978a,b) of Exeter University, authors of numerous scientific articles on the biology of this species.

Until recently there has been little critical assessment of the extent of damage caused by this exotic carnivore, and much of what has been said refers to anecdotal accounts of their alleged depredations on domestic livestock, and perpetuates the various misconceptions concerning their biology. Savage disruptive alien or well-integrated immigrant, the mink is here to stay. No amount of anger or enmity will empower us to eradicate it now. So it is time to cast off emotive reactions and come to understand the new balance being struck in our riparian community.

EXPLODING THE MINK MYTH

One of the most notable misconceptions relates to the mink's breeding biology, reputed (incorrectly!) to compare well with that of the rabbit. The often reported 'plagues of mink' are usually the result of over-zealous trapping effort drawing in mink from adjacent territories.

> Mink are multiplying tremendously, and if there is not a concerted effort, they will reach epidemic proportions. What we need is concerted action. I am surprised that conservationists will not help because mink are decimating flora and fauna as much as road-side spraying.
>
> *Daily Telegraph*, September 1982

A plague of mink is inconceivable. They are solitary, territorial animals whose intolerance of other mink will always ensure a low population density. Only at two stages in their life history do mink show any resemblance of social behaviour. Firstly, mink may consort for a limited period during the mating season. Their curiosity and wide-ranging behaviour in search of females leads to many animals being captured at this time giving the illusion of a large population. The second occasion, for another short period during July and August, occurs when the female may be seen with her litter of the year in attendance just prior to their dispersal. At this time of year, there is only a short period of darkness which she can use to provide an adequate supply of food for her rapidly growing litter. Consequently, hunting is often diurnal and the sight of three or four well-grown animals hunting a river bank or pool will undoubtably cause a river-keeper's heart to palpitate. Naive and curious, dispersing mink are easily caught, again giving the impression of 'plague' proportions. Mink regulate their own numbers according to the availability of suitable habitat containing adequate prey through the mechanism of territoriality. There does not have to be a natural predator, their aggressiveness to members of their own kind is more than adequate to control the population.

Newspaper reports (see Fig. 11.1) concerning mink activity show a good chronological correlation with these significant events in the mink's year

Exploding the Mink Myth

FIGURE 11.1 *Newspaper reports of the mink menace.*

which affect its visibility to man more than its level of predation. Every spring and late summer the mink biologist must steel himself to meet the wave of ill-conceived publicity regarding the status of the feral population.

A major cause of the misconceptions is due, as Birks (1990) points out, to the 'scapegoat effect', whereby the media demand simplified answers to complicated ecological questions. The mink's true story does not translate well into journalese, and mostly it is not what the readership wants to read. Something or someone must take the blame, and who better than the mink.

To evaluate the pest status of the mink, we need to assess firstly the extent of any significant economic loss or damage to domestic stock, the impact on native wildlife through predation or competition, and the nuisance value of mink to the general public. Secondly, so that a policy for the management of feral mink can be suggested, we need to assess the numbers, fecundity and dispersal capabilities and the potential for their control.

Two additional points need to be considered when assessing pest status from dietary studies. Firstly, population studies on prey have tended to be carried out retrospectively, usually some time after the presence of mink has been determined. Secondly, most studies have averaged the dietary information across individual mink in the population, and over a long period. This may have the effect of masking depredations which occur in the short term or by particular individuals.

PREDATION ON NATIVE FAUNA

The results from dietary studies have already been presented. We need only consider here those prey species which have been identified as being potentially at risk from mink, such as ground-nesting birds (ducks, moorhens, coots and seabirds), or have some commercial value (poultry, gamebirds and salmonid fishes).

Mink and salmonids

Concern is frequently expressed over the effect of mink on salmonid stocks (e.g. Lever 1985), but it is unlikely they do more than check the surplus in those places and seasons of abundance. It should be pointed out that the daily intake of mink is small relative to the 1–1.5 kg of fish needed to sustain an otter (Erlinge 1968). Otters, fish and fishermen seem to have co-existed happily enough in the past. By comparison, the mink is an opportunistic fisher, and not well adapted for swimming and underwater pursuit (see Chapter 4). The dietary studies in which predation on salmonids was found to be important, typically oligotrophic rivers, tend to be in areas where alternative prey were uncommon (Akande 1972, Cuthbert 1979, Chanin & Linn 1980). If the fish themselves become scarce, the mink is more likely to redirect its hunting efforts towards terrestrial prey.

Little supportive evidence has been found for any deleterious effect on fish stocks and angling interests. Chanin & Linn (1980) examined records of commercial salmonid net fisheries in the Teign estuary (one of the first rivers to be colonized by feral mink, and subsequently subjected to considerable research effort). From 1951 to 1968 migratory sea-trout catches increased four-fold, while salmon catches changed little over the same period. A decline in the numbers of salmonids thereafter was precipitated by salmon disease (UDN, ulcerative dermal necrosis). No changes in the catch by netsmen or rod fishermen could be attributed to the increasing mink population. Linn & Chanin (1978a) rightly comment on the confusion in the popular press between killing of individuals and the depletion of populations. Fish stocks have a remarkable ability to respond to depletion because of their great fecundity. Pressure on the scale imposed by a relatively small number of mink is likely to be negligible. On the upper reaches of the oligotrophic River Teign, brown trout were too numerous for the available food supply. The mink's cropping of the resident brown trout population may be beneficial to fisheries interests by allowing the remainder to grow to a greater size. The only coarse fish species recorded in any quantity in the diet of mink has been the eel (*Anguilla anguilla*) at 11–23% (Cuthbert 1979, Wise *et al.* 1981). The removal of eels, a predator of fish eggs and a favoured food of mink, may only serve to improve game fish stocks.

Mink and waterfowl

Ground-nesting birds such as waterfowl have been shown to be at risk during their breeding season and are preyed on in marsh and wetland habitats in the USA (Sargeant *et al.* 1973, Eberhardt & Sargeant 1977); although here the situation was complicated by loss of nesting habitat which had the effect of concentrating the mink's food. In the British Isles, waterfowl do not appear to be as important. Yet these prey groups have also been repeatedly suggested (Lever 1977, 1978a,b, 1985) as being major victims. It should be added that ornamental ducks on a small pond, especially when pinioned and kept at high density in the vicinity of streams or rivers will be, not unexpectedly, particularly vulnerable. Artificial, high-density collections of exotic wildfowl, such as those exhibited by the Wildfowl and Wetlands Trust in England can suffer badly from the depredations of mink.

Inevitably waterfowl are more at risk from mink predation than any other group of prey. They are ground nesting and found precisely where the mink is likely to encounter them in its riparian wanderings. The brunt of the predation seems to fall upon moorhens, coots and ducks, and the British Trust for Ornithology has circumstantial evidence that their numbers have been suppressed on some mink-infested waters. The important fact is that the effect is local: nowhere in the UK have mink caused widespread population declines. Counts of overwintering waterfowl and wading bird species, coordinated by the Wildfowl and Wetlands Trust since the 1960s, show no signs of a national decline that can be attributed to mink predation; indeed most species of waterfowl have increased. In north-east Iceland it has been reported that the waterfowl population on Lakes Myvan and Vikingavatn had decreased from 50 000 in 1961 to 15 500 in 1974. Although this was originally attributed to mink, doubt has been cast on this aspersion in recent years.

What is undoubtedly true is that there are a wide range of reasons for the decline of many of our waterbird populations, including drainage of wetlands, riverbank clearance, pollution of rivers and estuaries and increased levels of disturbance resulting from increased leisure activities. Such habitats become sub-optimal, yet because of the paucity of alternative breeding sites, birds have nevertheless to use them, often breeding at higher density than is normal and so become more vulnerable to predation, not only by mink but also by other ground-based predators.

Mink and sea-birds

The mink's predation on colonial ground-nesting sea-birds appears to be one area where considerable damage could be caused. Some features that increase their vulnerability are the concentration of breeding activities for regional populations at a few, heavily used sites, poor defensive behaviour due to the previous lack of terrestrial predators, and the extent to which harassment can upset normal breeding behaviour resulting in reduced breeding success in addition to the numbers of direct kills.

A population of the black guillemot (*Cepphus grylle*) has been sharply reduced since 1976 following the arrival of mink on the Söerskär peninsula of Finland in the summer of 1974. The birds occupied an offshore island. In 1974 and the following year, damage attributed to mink was small, but in 1976, 42% of the breeding population of 371 pairs was lost. In 1977 no mink reached the island, but in 1978, 41% of the 328 nests were predated. In 1981 only 167 breeding pairs remained. Similar allegations have been made in

Iceland concerning losses of breeding sea-birds including shags, common gulls, common and arctic terns, black guillemots and puffins. In northern Norway the decapitation of kittiwake chicks has also been blamed on a marauding mink. In North America the entire reproductive output of one colony of spotted sandpipers (*Actitis macularia*) was wiped out in one incident (Oring *et al.* 1983).

In other instances, after an initial period of destabilization, some resolution has been reached. Mink have been reported to cause considerable damage at breeding colonies of common terns (*Sterna hirundo*) on small islands off the western coast of Scotland, with the result that breeding has failed, and in some cases, colonies deserted. The birds' response has been to re-establish their colonies on islands further offshore where, for the moment they are immune from the mink's attack (Craik quoted in Birks 1990). Between 1972 and 1974, Gerell found eider ducks and gulls in Sweden suffering severe predation from mink, causing them to desert their nests and to abandon subsequent breeding attempts. But in the following years a balance became established with evidence that the birds had learned to mob the predator (Gerell 1985).

A small offshore island became an interesting feature of our study area in southern Scotland (Dunstone & Birks 1983, 1985, Ireland 1990). The 10 ha island was used by a colony of herring gulls (*Larus argentatus*), but was also the exclusive territory of a female mink. Over the three breeding seasons we observed her activities, her food requirements and those of her kits before dispersal were met by 'predation' upon the gulls. Sea-bird carrion was plentiful on the island and was used by the mink, and a considerable quantity remained untouched by her at the end of the breeding season. In such a case, it seems that the mink's territoriality prevented heavy exploitation of the birds. Surplus killing would be unlikely by a mink habituated to colonial prey. At least with some sea-bird species, a healthy balance seems possible.

Mink and mammals

Large mammals are by far the most economic prey for the mink, and lagomorphs, particularly rabbits, are always taken if available.

In five of the ten studies conducted on the minks' diet in the UK, the most frequently taken item was lagomorph prey. Rabbits, when common in the study area, are the most heavily exploited mammal, and often the most important food overall for their bulk contribution to the diet (Jenkins & Harper 1980, Wise *et al.* 1981, Dunstone & Birks 1987).

Most mammal species taken by mink are abundant, and in the case of the rabbit and the rat, are pest species themselves, and unlikely to provoke concern. There are two species which have to be viewed in a different light. In Spain, feral mink are thought to be responsible for the decline of the European desman (*Galemys pyrenaicus*), a small insectivore resembling a large water shrew that inhabits tumbling mountain streams in the Pyrenees. While the species undoubtedly does fall prey to mink, populations of this animal are still found in areas where mink abound. A far greater threat to the desman is posed by the flooding of the river valleys for hydro-electric schemes. Here also, widespread habitat modification has led to the two species being forced to live in close proximity.

The second species for which the depredations of mink might pose a significant threat is the water vole (*Arvicola terrestris*). This prey approaches the optimum size for a mink and is readily accessible in the riparian niche. Populations of this semi-aquatic rodent have declined in the last decade over much of England and Wales. Female mink, at least, are small enough to enter the vole's burrows. However, only rarely have their remains been found in the scats of mink, and then at low frequency, (2–4% occurrence, Chanin & Linn 1980) even though both species live in the same habitat.

There can be little doubt that the mink has played a role in the later stages of the decline of water voles, but it is likely that habitat destruction, human disturbance and river pollution, particularly from organochlorine pesticides, have played the major part.

A full investigation of the status of the water vole is currently being undertaken in the UK. Most sites previously known to hold water voles still do, and in many mink are also present. There is evidence that the water vole may be adjusting to the presence of mink by showing a preference for small streams, brooks and ditches rather than the banks of larger rivers. Such areas would be visited less frequently by mink as they support a lower density of prey (Birks 1990).

PREDATION ON DOMESTIC STOCK

The occurrence of prey remains derived from domestic animals in the diet of mink is an extremely rare occurrence, even in those studies which have involved mink populations living in the vicinity of farm buildings and game-rearing pens (Dunstone & Birks 1987). Studies conducted over long periods and involving large samples of scats, show that poultry and game birds generally make up less than 1% of the diet; the highest recorded was 5.4% (e.g. Cuthbert 1979, Chanin & Linn 1980, Dunstone & Birks 1985). Initial fears about the importance of mink as an agricultural pest were exaggerated. Nevertheless, mink do take game birds; Fairley (1980) reports one female mink being shot in a pen of dead pheasants, and of feathers from game birds occurring in some mink stomachs. Similarly, Ward *et al.* (1986) report the occurrence of poultry feathers in 0.2% of their scat sample. Birks & Dunstone (1985) also report the occurrence of single partridge (*Perdix perdix*) and pheasant (*Phasianus colchicus*) carcasses in the dens of mink. Nevertheless the predation of mink on Galliforme prey species is often presumptive.

The most disturbing reports of stock damage involve mass killings of confined animals. Many carnivores exhibit the phenomenon of surplus killing under conditions where prey are encountered at high density and cannot escape from the predator. The phenomenon is not unique to mink, but tends to receive considerable publicity when it does occur. Surplus killing incidents are unusual in the wild, but have been reported from North America (e.g. Errington 1961). Here it is interpreted as an adaptive trait, conferring an advantage upon opportunistic predators which exploit food resources of varying abundance and availability.

Recently, Harrison & Symes (1989) have attempted a rather different approach to assess the economic pest status of mink. During the periods of survey, detailed records have been kept by the UK Government's Ministry of Agriculture, Fisheries and Food of the reports from landowners and farmers in south-west England of the damage attributed to mink. Although not exhaustive or particularly rigorous, such an analysis can provide useful information. The surveys include some 108 reports of damage during the period 1961–1970, and another 87 during 1985 and 1986.

The killing of poultry and domestic waterfowl amounted to some 60% of incidents reported (Fig. 11.2). Of these most involved small numbers, from

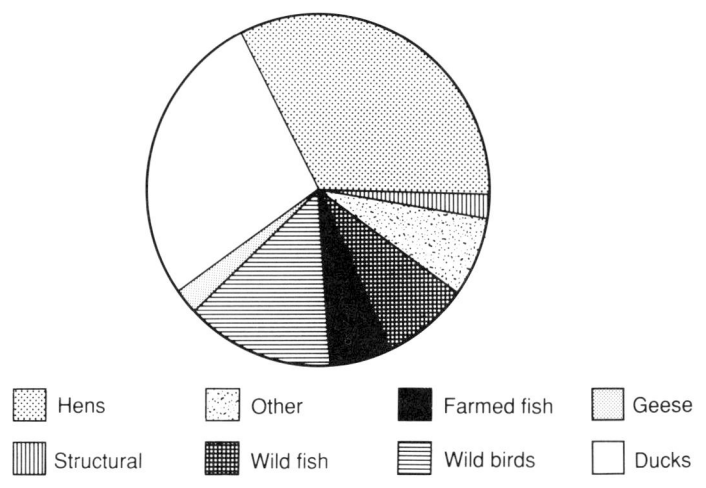

FIGURE 11.2 *Frequencies of different types of damage caused by mink in south-west England (1985–1986). Source: Harrison & Symes (1989).*

one to five kills, but exceptionally as many as 40 chickens were killed in a single incident (Fig. 11.3). Mink predation on poultry occurs with a higher frequency than that by foxes, badgers or other mustelids. However, it seems that incidents attributed to mink are more likely to be reported. Given that an individual laying hen is worth about £2, individual incidents are relatively trivial in terms of economic damage. Most UK poultry production is intensive, and takes place in closed buildings where, if they are adequately maintained, the birds should be safe from the depredations of mink. The recent public demand for free-range products is likely to lead to increased vulnerability to mink predation.

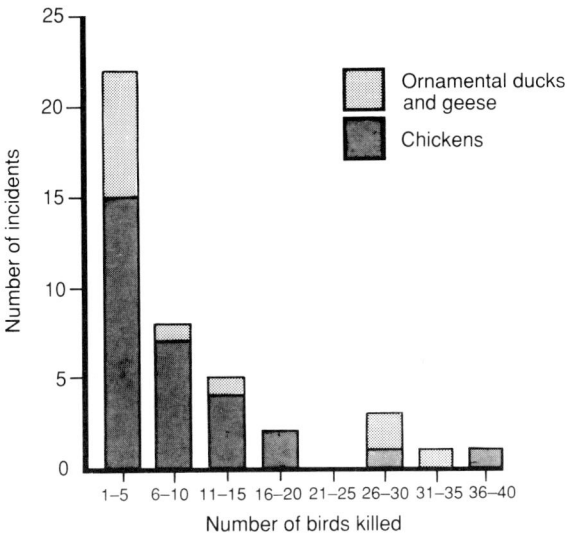

FIGURE 11.3 *The number of hens and ornamental waterfowl killed during mink attacks. Source: Harrison & Symes (1989).*

MAFF also recorded a number of kills of young pheasants kept in release pens, some involving in excess of 100 individuals. Similar instances have been reported in America, particularly in situations involving the artificial, usually high-density, propagation of wildfowl.

Predation at trout farms

Given the aquatic habits of mink, the high incidence of fish in the diet of native and feral animals, and the fact that fish farms are by necessity located close to waterways, it is not surprising to find that such enterprises receive frequent visits from marauding mink. Harrison & Symes (1989) conducted a questionnaire-based survey of a sample of 75 fish farms. Of those contacted, 46% had experienced visits by mink. However, as far as the farmers were concerned herons (*Ardea cinerea*) posed a far more serious predator problem. The type of damage attributable to mink ranged from direct predation and damage to fish, to damage caused to underwater holding-cages as the mink attempted to gain access to the confined prey. Fish have also been reported to die from stress or panic during attempted capture by mink. Damaged fish, caused by bites and scratches, were more likely to succumb to fungal infection. The size of trout taken varies from small fingerlings to fish up to 0.5 kg. One farmer claimed to have lost 500 trout of 50–75 g in just 1 month; in the following month he trapped 40 mink! This incident occurred during August when juvenile mink would be dispersing. Such incidents should be rarer at other times of the year when the farm would lie within the territory of one animal whose food requirements will be considerably less.

INTERACTION WITH NATIVE CARNIVORES

Competition between mink and otters

Competition occurs where two species are using a resource, such as food type or den, which is actually or potentially limiting. There is little reason to believe that feral populations of mink should not successfully co-exist with the otter. This is after all the case of native American mink with the river otter (*Lutra canadensis*) in the USA and Canada, and the European mink with the European otter in Russia and parts of Europe. Complete competitive exclusion is evidently not occurring.

In the case of competition for food between the mink and otter, the shared prey resource will usually be fish, of which otters are the considerably superior predator in terms of hunting efficiency. The otter is a fish specialist, whereas the mink is a generalist and will take terrestrial prey if it is more readily available. If competition for fish became intense, the mink would be the first to back out.

North America Observations of sympatric river otter and mink populations in North America have revealed differences in their foraging behaviour. Otters usually forage from the water for their prey, even if it is terrestrial in origin. In contrast, mink often forage for terrestrial prey while travelling along the shore. Aquatic prey are usually taken by diving into the water after sighting it from an out of water vantage point (Melquist *et al.* 1981).

Analysis of 657 mink scats and 1902 river otter spraints from west-central Idaho showed fish were the commonest prey for both predators (mink 59%; otters 97%), but whereas mammals, birds and invertebrates were also important prey for mink, these were of negligible significance to the otter. The predominance of fish in both diets suggests a high degree of overlap, but this was belied by closer examination of the prey (Fig. 11.4). The greater body size

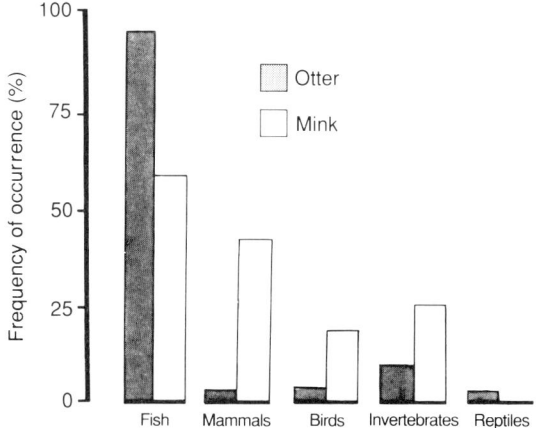

FIGURE 11.4 *Major prey categories in the diet of mink and otter in west-central Idaho. Source: Melquist et al. (1981).*

and better adaptations for underwater hunting possessed by the otter allowed them to forage on a more extensive range of generally larger fish than did mink. For example, large-scale sculpins (*Catostomus macrocheilus*) of 35–45 cm were commonly taken by otters but not predated by mink. Mottled sculpins (*Cottus bairdi*) were thought to be less available to mink because of their bottom-dwelling habits. The greatest amount of overlap occurred in the use of cyprinid fishes, redside shiners (*Richardsonius balteatus*) and speckled dace (*Rhinichthys osculus*) 7–12 cm long. Kokanee (*Oncorhynchus nerka*), a migratory salmonid, were of considerable importance to otters (43% of fish taken) during spawning runs, but were infrequently taken by mink, at least early in the season. Later, after they had spawned, spent fish were taken by mink in shallow water or as carrion on the shore.

Europe Estimates of dietary overlap of 40% between otter and feral mink have been demonstrated on a eutrophic lake and a moorland river in south-west England (Wise *et al.* 1981). In contrast to North American studies, otters and mink showed a lack of size selection when preying on fish. Fish constitute the bulk of the otter's diet all year round, but fish predation by mink was seasonal. Thus dietary overlap was greatest in autumn and winter, when fish were easier to catch in cold water. (Fig. 11.5). Greatest competition might be

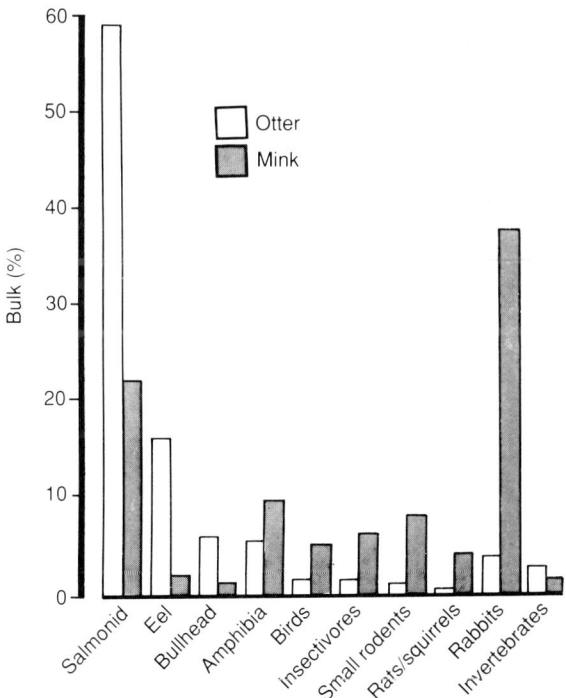

FIGURE 11.5 *Major prey categories in the diet of mink and otter in south-west England. Source: Wise* et al. *(1981).*

expected to occur during the summer months when fish were less available. At this time it is generally noted that mink are preying heavily upon rabbits, if available locally. In those areas where otter diet was studied before the introduction of mink there has been no subsequent change recorded in the diet of the otter (Chanin 1986) suggesting that competition is not occurring.

On a Swedish river, dietary overlap ranged from 50% in summer to 70% in winter when some foods became unavailable because of ice cover (see Fig. 11.6), Erlinge 1972). On Lake Flisbysjön there was a lower density of mink than would be expected, possibly due to interference from the large population of otters present. During winter the ice cover on lakes and rivers, which may last for 3–5 months, substantially reduces the number of aquatic areas that can be hunted by either predator. This occurs at a time when fish prey are a main dietary item for both predators. Competition for prey would appear to be strong in this case, and it is significant that the mink is the species more limited by it.

In summary, dietary overlap between mink and otter in the use of fish prey will not be critical unless the resource is limiting, as it might be in a sub-optimal habitat, for example impoverished oligotrophic streams. Even here it is likely that the mink will suffer more than the otter because it is the poorer

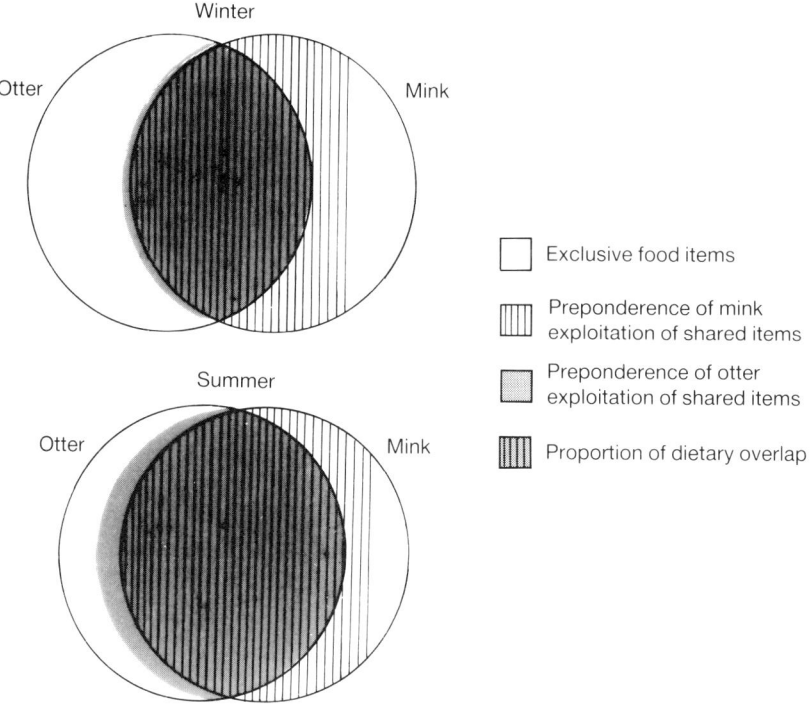

FIGURE 11.6 *Dietary overlap between the mink and otter: top, winter; bottom, summer. Source: Erlinge (1972).*

aquatic hunter. If available, mink will then switch to terrestrial prey. In some studies competition was avoided because the different sized predators took differing prey species. This again works to the advantage of the otter, since it can take both small and large prey, while only smaller fishes can be taken by mink.

Concern that mink may prey on otter cubs seems to be without foundation. Indeed, Grigor'ev & Egorov (1969) reported that the otter in Russia 'is a serious enemy' of the mink. Analysis of 880 otter spraints revealed remains of six mink. Similarly, Novikov (1956) reports the otter to 'vigorously hunt mink' in Russia.

Since the two species inhabit similar aquatic habitats and are both secretive in nature, spending long periods lying up in dens, it might be expected that there would be competition between them for suitably secure sites, especially in areas where these are in short supply.

In North America it has been reported that both species will use the same log jams as den sites (Melquist *et al.* 1981). Mink, because of their considerably smaller size, were able to live in places inaccessible to otters. Mink used a variety of structures including log jams, stream bank, brush and debris, rock crevices and beaver lodges, whereas otters only used the first two types. The use of log-jam dens by mink peaked in December when use by the otter was at its lowest, but it was not thought that this was a response to the absence of the otter.

Simultaneous radio-tracking of mink and otter was achieved on 65 days, mink and otter were observed to be foraging at the same time in the same log jam on 18 of those days. Although mink appeared to be using drier parts of the log jam while the otter foraged amongst logs in the main river channel, there were frequent occasions when the two species were recorded active within 5 m of each other. During 21 days of monitoring, mink were active at times when otter were resting in the same log jam or vice versa, often moving in close proximity without disturbing the other. The two dens most frequently used by otter and mink were within 3 m of one another. To what extent the resource partitioning observed in this study is due to the presence of a unique and massive log jam is not known.

In Britain also, mink have been found occupying the same holt site as the otter. Indeed, on one occasion a mink hunt flushed a mink and an otter from the same stickpile on the River Creedy in Devon (Green *et al.* 1986).

Decline of the otter
That mink were blamed for the decline in otter populations in the 1960s and 1970s is not surprising. The vicious alien predator was fast spreading along our rivers at a time when the much loved otter appeared to be in equally fast decline. From the subsequent studies outlined above, there is little evidence to support the contention that mink have adversely affected the status of otters. The view is, however, well entrenched in mink mythology (Lever 1985) and frequently one reads that 'as the mink moved in, the otters gradually disappeared'.

A number of causes have been considered for the decline of the otter, including mink (Chanin & Jefferies 1978). The abundance of otters, as estimated from the hunting records of otter hound packs, generally showed an increasing population up until 1956. The fact that the decline took place so quickly and simultaneously across the country, led to a feeling that it must have been brought about by a change in circumstances occurring in about 1957.

Hunting itself has been suggested as the cause, but analysis of records show a remarkably constant rate of killing, up until a voluntary ban was imposed by the Masters of Otter Hounds Association when they became aware of the severity of the problem. Although deaths as a result of hunting did not cause the decline there was evidence to suggest that in areas of a large cull, the decline in the otter population was greater and undoubtedly led to a greater pressure on the otter. In many countries riparian habitat destruction, particularly for enhanced drainage schemes, and increased disturbance brought about by an increase in leisure activities (fishing and boating) may have been involved in the decline. However, there was no evidence to suggest any difference in these activities before and after 1957.

The timing of the otter's decline however, links it more closely with the introduction in 1955 of organochlorine insecticides, dieldrin and aldrin, for use in sheep dip and as seed dressings. Many of the uses of these pesticides in agriculture were banned in the early to mid 1960s when the devastating effects of pollutants on the aquatic food chain were realized.

This insidious catastrophe was an unfortunate coincidence as far as the mink was concerned. By the time the decline of the otter had been brought to the attention of the general public, the mink was widespread and its alleged notoriety documented in the press. A convenient scapegoat was at hand. In reality, the mink was absent from many of the places where the decline of the otter was being noted. Even now, many years after the event, the view is often expressed in sporting and country magazines that the mink was a significant cause of the otters' problem.

> I have found that, over the years, as mink moved in, the otters gradually disappeared.
> Having spent most of my working life in the country and by the rivers as a gamekeeper, I actually saw the killing of a young otter by a wild mink'.
> *Western Morning News* (17 February 1981)

Unlike populations of predatory birds that were similarly affected by these toxic chemicals, the otter populations have been slow to rebuild. This is possibly because their populations became too fragmented, perhaps exacerbated by hunting pressure, and movement between isolated groups was prevented because of the lack of cover caused by habitat destruction. Even when it was accepted that the mink was not to blame for the decline of the otter, doubt was then expressed as to whether their presence along a river would in any way hinder the ability of the otter population to re-establish itself. Recently, partly through a programme of captive breeding and re-introduction, populations of otters are recovering in England. They are now present on many rivers that contain mink populations. The low otter population may have facilitated the rapid spread of mink, but it is more likely that the mink will give way to the returning otter, rather than impair its recovery. Indeed, Birks (1990) reports evidence that in some areas where otters have made a significant recovery over the last 10 years, mink populations are now lower. The causal relationship has yet to be investigated; what is certain is that the mink poses no direct threat to the otter.

A problem arises with the continued use of hounds to hunt mink. In view of the tendancy for mink and otter to be found on the same rivers and occasionally to use the same resting places, this will inevitably cause disturbance to the otter and could represent a significant threat in areas where they are scarce or attempting to re-colonize.

Mink and other mustelids

The potential interaction between American and European mink has already been discussed. In Europe, the only other carnivores with which the mink is likely to be in direct competition are the polecat (*Mustela putorius*) and the stoat (*Mustela erminea*). The staple item of diet for these predators is the rabbit (Day 1968, Walton 1968). Populations of rabbits are such that competition with mink is unlikely to be a limiting factor affecting the status of either native carnivore. Furthermore, the polecat appears to be extending its range even in areas where mink are present (Birks 1986). This may be due partly to the minks' association with waterways, on which the polecat appears to be less dependent.

THE VERDICT

Early worries concerning the impact of mink as a significant predator of native fauna and domestic stock, or as a competitor with native carnivores are largely without foundation. It can be seen that the relatively small size of mink, and hence their low energy requirements, coupled with their low population density, means that they have little impact overall. General conclusions from dietary studies indicate that mink are not specialist predators; they exist in a flexible balance with prey and do not appear to have threatened the viability of any one prey species. In Britain there have not been any extinctions or drastic reductions of any species of prey that can be directly and unequivocably attributed to mink.

The major problem caused by mink predation concerns their effect on native bird species. While distressing incidences of killings may occur, in most cases the threat to populations is slight, unless habitat damage has already put the population in a precarious position. Here it must be remembered that mink are territorial, and the size of the territory is influenced by two factors—the habitat quality and the presence or absence of other mink. Because of this spacing mechanism, any local abundance of food or seasonal glut will only be available to the territory holders, thus allowing only a small number of animals to share the food, and preventing mink from surrounding areas gaining access. Thus, for example, on Ross Island where there were abundant ground-nesting herring gulls available for most of their breeding season, only one female mink preyed on the colony because it lay within her territory. The colony has continued to expand despite the annual predation pressure from mink (Birks 1986).

Considerable concern has been expressed about the predation by mink of domestic stock, particularly chickens, game birds and fish. Undoubtedly mink can pose a serious threat to agricultural holdings of high-density stock, for example, game-bird rearing pens, chicken farms and fish rearing ponds, as can many other predators such as the fox, otter, dog and cat. Mink are frequently blamed for the surplus killing of domestic prey, usually when encountered in confined conditions. These incidents can be guarded against by good husbandry, which will additionally provide protection against native predators. The maintenance of traps in the vicinity of pens is all that is usually required to control mink in these instances.

In terms of direct losses, the economic impact across the country of mink as a predator of domestic stock is almost negligible, although the individual farmer who has just lost 50 of his best free-ranging hens down by the river might not see it that way. But there are hidden costs associated with the presence of this alien predator. The cost of control operations both in trapping equipment and time can be considerable. The unsuccessful eradication campaign of the Ministry of Agriculture in a 5-year trapping programme cost £105 000 between 1965 and 1970. If such a campaign were even to be contemplated today it would cost many millions of pounds. The individual farmer may spend an appreciable sum of money mink-proofing his stock, but the benefits derived by also excluding other predators may be substantial. In some locations, the prevalence of mink predation may threaten the viability of certain farming operations. Such restrictions in the siting of these enterprises may also be a cost. Before the presence of feral mink, most farms kept a few free-range chickens or small flocks in deep-litter houses. Today the poultry industry is intensive with far fewer but much larger flocks kept in battery units where they are immune from attacks by predators.

THE EFFICACY OF CONTROL OPERATIONS

TRAPPING

The cost involved in mounting 'effective' control operations can be substantial. The job is magnified by the presence of transient individuals who will rapidly replace resident territory holders that are removed. Consequently many more animals will be trapped on a given stretch of river than were initially present, and recolonization from adjacent areas is almost inevitable. In Sweden, considerable effort was expended from the 1940s to the 1960s by the Sportsmen's Association in trapping mink in an attempt to protect their hunting interests. Their annual catch is shown in Fig. 11.7.

Successful attempts to trap out mink populations have only occasionally been documented. Most fail or have only a limited effectiveness until mink re-invade. One such effort was made on the River Otter in Devon in 1973 (Chanin 1981). A total of 47 mink from 27 km of river were killed between April and September. It is unlikely that all mink were captured, and the area was recolonized during the next breeding season. As Chanin (1981) points out, even if mink had been eradicated on this river they would inevitably have re-invaded from adjacent rivers. The cost of establishing a 'mink-free zone' is not warranted in view of the small amount of economic damage they do.

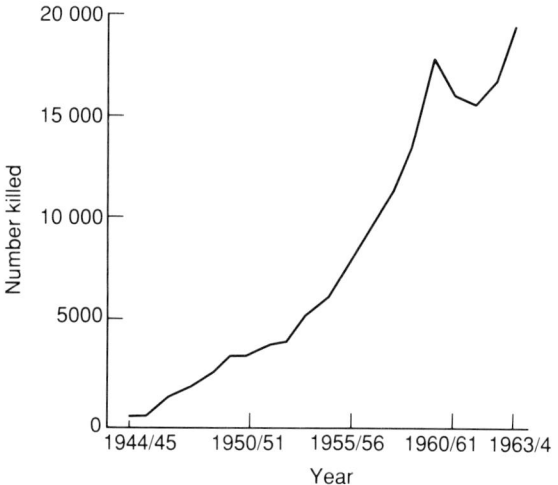

FIGURE 11.7 *Total catch of mink in Sweden from 1944 to 1964. Source: Gerell (1967a).*

HUNTING

Otter hunting with packs of hounds was banned in the UK in 1975. At least six of these packs subsequently switched to hunting mink. Other packs were formed with the specific purpose of hunting mink. By December 1984, 20 such packs had registered with the Masters of Minkhounds Association.

Hunting is carried out from April to September, a period which includes the mink's breeding season. Den sites are located by the dogs as they search the riverbank, and an attempt is made to bolt the quarry which then dashes to its next refuge. There have been a considerable number of reports of damage to river banks and their vegetation caused by hunt servants in attempting to dislodge mink from their dens. Although precise figures are not available each hunt probably accounts for some 40–50 mink per season, amounting to an annual nationwide slaughter of 700–800 (Birks 1986).

Only one study (Birks 1981, 1986, 1989) has addressed the effectiveness of hunting with dogs as a means of controlling mink. In an analysis of hunting records of the Cornwall and Devon Minkhounds (south-west England) from 1976 to 1980, the pack hunted on 156 days during which 84 mink were caught. Two-thirds of the mink located by the hounds successfully evaded capture.

On one occasion, the minkhounds hunted through Birks' study area, at which time five or six resident mink were known to be present. Three of these residents were found by the hounds, two of which escaped into secure rocky dens, only one was killed. Although this seems a remarkably inefficient means of control, the extent of persecution may be more subtle. The recently born litter of the female were left to perish. In another case, a presumably pregnant female hunted in March failed to produce a litter that year. On mink farms interference causing stress at such an early stage in pregnancy is likely to lead to abortion.

APPROPRIATE CONTROL

It is doubtful if the agricultural or ecological impact of mink on domestic stock or native prey species is sufficient to warrant expenditure on widespread control measures.

The most cost-effective way of guarding against mink attack is not by control, but by excluding the predator. Mink are small predators and can quite easily gain access to poultry enclosures and sheds through rat holes, possibly when hunting the rats themselves. Particularly valuable collections of exotic waterfowl can be protected from mink by means of a perimeter wire fence with metal baffle at the top to prevent the mink climbing in. Preventative trapping should be carried out at times of the year when it is likely to have most effect, for example, by removing pregnant females in spring and territorial residents in autumn after dispersal movements have largely finished.

While mink must be accepted as endemic across much of northern Europe, its further spread into ecologically or agriculturally sensitive areas should be prevented. In 1977 permission was sought, but refused, to establish a mink farm on the island of Westray, Orkney, Scotland. This island supports one of the largest colonies of breeding sea-birds in the British Isles, amounting to some quarter of a million birds of 19 species. A small endemic rodent, the Orkney vole (*Microtus arvalis orcadensis*) would also have been vulnerable. The Mink (Keeping) Order 1987 prohibits the keeping of mink on any offshore island of Great Britain other than those where they are farmed or already established in the wild.

LEGISLATION PERTAINING TO MINK

Made under the Destructive Imported Animals Act 1932, the Mink (Keeping) Order 1987 prohibits the keeping of mink within Great Britain except under licence, and requires occupiers to notify the presence of (unlicenced) mink on their land. The keeping of mink is prohibited on any offshore island of Great Britain other than the Isle of Wight, the Isles of Arran (and nearby Holy Island), Harris and Lewis and its associated islands north of a line dividing Harris from the islands of Shillay, Pabbay, Killegray, Langay and Gilsay. The keeping of mink in the Caithness and Sutherland districts of the Highland region of Scotland is also forbidden.

Licence fees and keeping conditions are specified in the Keeping Regulations and penalties are proscribed under the Act. Licence fees, £115, in 1992, are kept under review and have to reflect the administration and inspection costs involved.

In addition, the Wildlife and Countryside Act 1981 makes it an offence to release mink or to allow them to escape into the wild. Importation, except under licence, is prohibited by the Rabies (Importation of Dogs, Cats and Other Mammals) Order 1974, as amended.

GENERAL OVERVIEW

The popular view is of an out-of-control bloodthirsty predator which rampages along our riverbanks and wreaks havoc on a grand scale, leading to major losses of fish stocks and waterside wildlife. The alternative view is that the mink is occupying a previously vacant niche, with the result that its permanent impact upon our fauna will be relatively slight.

Failure to tease the myth from the reality could have dire consequences. Where the mink's innocence is falsely protested in response to an ecological problem, then possibly avoidable damage may occur through inaction or through tackling the wrong cause. Conversely, where the mink is falsely blamed, the genuine causes of wildlife problems may be ignored with serious consequences. In some instances 'the cure may be worse than the disease', as traps set to control mink may ensnare other wildlife.

References

AGRICOLA, G. (1549) *'De Animantibus subterraneis liber'*, Basil.
AKANDE, M. (1972) Food of the feral mink (*Mustela vison*) in Scotland. *J. Zool. Lond.* **167**: 475–479.
ALLEN, A.W. (1983) Habitat suitability index models: mink. *US Dept Int. Fish. Game and Wld. Ser.* OBS 82/10.61.
APFELBACH, R. (1973) Olfactory sign stimulus for prey selection in polecats (*Putorius putorius* L.). *Z. Tierpsychol.* **33**: 270–273.
ARNOLD, T.W. & FRITZELL, E.K. (1987) Food habits of prairie mink during the waterfowl breeding season. *Can. J. Zool.* **65**: 2322–2324.
ATKESON, T.Z. (1956) The incidence of crippling loss in steel trapping. *J. Wldl. Mgmt.*
AULERICH, R.J. & SWINDLER, D.R. (1968) The dentition of the mink (*Mustela vison*). *J. Mammal* **49**: 488–494.
AULERICH, R.J., RINGER, R.L. & IWAMOTO, S. (1974) Effects of dietary mercury on mink. *Acta Environ. Contam. Toxicol.* **2**: 43–51.
BAINBRIDGE, R. (1958) The speed of swimming of fish is related to size and to the frequency and amplitude of the tail beat. *J. Exp. Biol.* **35**: 109–133.
BALLIET, R.F. & SCHUSTERMAN, R.J. (1971) Underwater and visual aerial acuity in the Asian 'clawless' otter *Amblonyx cineria cineria*. *Nature Lond.* **234**: 305–306.
BEVANGER, K. & ÅLBU, O. (1986a) Mink *Mustela vison* i Norge. *Okoforsk Utredn.* **6**: 1–73.
BEVANGER, K. & ÅLBU, O. (1986b) Decrease in a Norweigan feral mink *Mustela vison* population—a response to acid precipitation? *Biol. Conserv.* **38**: 75–78.
BIRKS, J.D.S. (1981) Home range and territorial behaviour of the feral mink (*Mustela vison* Schreber) in Devon. Ph.D. Thesis. Exeter University.
BIRKS, J.D.S. (1986) *Mink*. Oswestry, UK: Anthony Nelson.
BIRKS, J.D.S. (1989) What regulates the numbers of feral mink? *Nature in Devon* **10**: 45–61.
BIRKS, J.D.S. & DUNSTONE, N. (1984) A note on prey remains collected from the dens of a coast-living mink (*Mustela vison*) population. *J. Zool. Lond.* **203**: 279–281.

BIRKS, J.D.S. & DUNSTONE, N. (1985) Sex-related differences in the diet of mink (*Mustela vison*). *Holarctic Ecol.* **8**: 45–52.
BIRKS, J.D.S. & LINN, I.J. (1982) Studies on the home range of feral mink (*Mustela vison*). *Symp. Zool. Soc. Lond.* **49**: 231–257.
BIRNEY, E.C. & FLEHARTY, E.D. (1968) Age and sex comparisons of wild mink. *Trans. Kansas Acad. Sci.* **69**: 139–145.
BLEAVINS, M.R. & AULERICH, R.J. (1982) Feed consumption and food passage time in mink and European ferrets. *Lab. Anim. Sci.* **31**: 268–269.
BRINCK, C., GERELL, R. & ODHAM, G. (1978) Anal pouch secretion in mink *Mustela vison*. *Oikos* **30**: 68–75.
BRINCK, C., ERLINGE, S. & SANDELL, M. (1983) Anal sac secretion in mustelids. *J. Chem. Ecol.* **9**: 727–744.
BROWN, J.H. & LASIEWSKI, R.C. (1972) Metabolism of weasels: the cost of being long and thin. *Ecology* **53**: 939–943.
BURGESS, S.A. (1978) Aspects of mink (*Mustela vison*) ecology in the southern Laurentians of Quebec. MSc. Thesis. McGill University.
BURGESS, S.A. & BIDER, J.R. (1980) Effects of trout stream improvements on invertebrates, trout populations and mink activity. *J. Wildl. Mgmt* **44**: 871–880.
BURT, W.H. (1943) Territoriality and home range concepts as applied to mammals. *J. Mammal.* **24**: 346–352.
BUSKIRK, S.W. & LINDSTEDT, S.L. (1989) Sex biases in trapped samples of Mustelidae. *J. Mammal.* **70**: 88–97.
CHANIN, P.R.F. (1976) The ecology of feral mink (*Mustela vison* Schreber) in Devon. Ph.D. Thesis. Exeter University.
CHANIN, P.R.F. (1981) Diet of the otter (*Lutra lutra*) in relation to feral mink (*Mustela vison*) in two areas of south-west England. *Acta Theriol.* **26**: 83–95.
CHANIN, P.R.F. (1983) Observations on two populations of feral mink in Devon. *Mammalia* **47**: 463–476.
CHANIN, P.R.F. (1986) *The Natural History of Otters*. London: Croom Helm.
CHANIN, P.R.F. & JEFFERIES, D.J. (1978) The decline of the otter *Lutra lutra* in Britain: an analysis of hunting records and discussion of causes. *Biol. J. Linn. Soc.* **10**: 305–328.
CHANIN, P.R.F. & LINN, I.J. (1980) The diet of the feral mink (*Mustela vison*) in southwest Britain. *J. Zool. Lond.* **192**: 205–223.
CLARK, S.P. (1970) Field experience of feral mink in Yorkshire and Lancashire. *Mamm. Rev.* **1**: 1–8.
CLEMENTS, F.A. & DUNSTONE, N. (1979) The threshold for high speed directional motion perception in the mink *Mustela vison* Schreber. *Anim. Behav.* **27**: 613–620.
CONROY, J.W.H. & JENKINS, D. (1986) Ecology of otters in northern Scotland. VI. Diving times and hunting success of otters (*Lutra lutra*) at Dinnet lochs, Aberdeen and in Yell Sound, Shetland. *J. Zool. London:* **209**: 341–346.
CORBET, G.B. (1966) *The Terrestrial Mammals of Western Europe*. Philadelphia.
CORBET, G.B. (1980) *The Mammals of Britain and Europe*. London: Collins.
COWAN, I. McT., WOOD, A.J. & KITTS, W.D. (1957) Feed requirements of

deer, beaver, bear and mink for growth and maintenance. *Trans. 22nd North Amer. Wild. Conf.* **22**: 179–188.

CUTHBERT, J.H. (1979) Food studies of feral mink (*Mustela vison*) in Scotland. *Fisheries Mgmt* **10**: 17–25.

DAGG, A.I. (1973) Gaits in mammals. *Mamm. Rev.* **3**: 135–154.

DANILOV, P.I. & TUMANOV, I.L. (1976) Mustelids of the north western USSR. 256 pp. (In Russian) Leningrad.

DAVIES, S.W. (1988) An investigation of the effects of various environmental parameters on the underwater foraging behaviour of the American mink *Mustela vison* Schreber. Ph.D. Thesis. University of Durham.

DAY, M.G. (1968) Food habits of stoats (*Mustela erminea*) and weasels (*Mustela nivalis*). *J. Zool. Lond.* **155**: 485–497.

DAY, M.G. & LINN, I.J. (1972) Notes on the food of feral mink (*Mustela vison*). *J. Zool. Lond.* **167**: 463–473.

DEANE, C.D. & O'GORMAN, F. (1969) The spread of feral mink in Ireland. *Irish Nat. J.* **16**: 198–202.

DUBIN, M.W. & TURNER, L. (1977) Anatomy of the retina of the mink. *J. Comp. Neurol.* **173**: 275–288.

DUNSTONE, N. (1978) The fish-catching strategy of the mink (*Mustela vison*): time-budgeting of hunting effort? *Behaviour* **67**: 157–177.

DUNSTONE, N. (1981) Swimming and diving behaviour of the mink. *Carnivore* **2**: 56–61.

DUNSTONE, N. (1986) Exploited animals—the mink. *The Biologist* **33**: 69–75.

DUNSTONE, N. & BIRKS, J.D.S. (1983) Activity budget and habitat usage by coast-living mink (*Mustela vison*). *Acta Zool. Fenn.* **174**: 189–191.

DUNSTONE, N. & BIRKS, J.D.S. (1985) The comparative ecology of coastal, riverine and lacustrine mink *Mustela vison* in Britain. *Z. Ange. Zool.* **72**: 52–70.

DUNSTONE, N. & BIRKS, J.D.S. (1987) The feeding ecology of mink (*Mustela vison*) in a coastal habitat. *J. Zool. Lond.* **212**: 69–83.

DUNSTONE, N. & IRELAND, M. (1989) The mink menace? A reappraisal. In *Mammals as Pests*, R. Putman (ed.) pp 225–242 London: Chapman and Hall.

DUNSTONE, N. & O'CONNOR, R.J. (1979a) Optimal foraging in an amphibious mammal. I. The aqualung effect. *Anim. Behav.* **27**: 1182–1194.

DUNSTONE, N. & O'CONNOR, R.J. (1979b) Optimal foraging in an amphibious mammal: II. an analysis using principal components analysis. *Anim. Behav.* **27**: 1195–1205.

DUNSTONE, N. & SINCLAIR, W. (1978) Comparative aerial and underwater visual acuity of the mink (*Mustela vison* Schreber). *Anim. Behav.* **26**: 14–21.

EAGLE, T.C. & WHITMAN, J.S. (1984) Mink. In *Wild Furbearer Management and Conservation in North America*, M. Novak, J.A. Baker, M.E. Obbard & B. Malloch (eds): Public Ministry of Natural Resources Ontario.

EBERHARDT, R.T. (1973) Some aspects of mink–waterfowl relationships on prairie wetlands. *Prairie Naturalist* **5**: 17–19.

EBERHARDT, R.T. & SARGEANT, A.B. (1977) Mink predation on prairie marshes during waterfowl breeding season. In *Proceedings of the 1975 Preda-*

tor Symposium, R.L. Phillips & C. Jonkel (eds): Montana Forest and Conservation Experimental Station University of Montana, Missoula.

EIBL-EIBESFELDT, I. (1956) Angeborenes und Erworbenes in der technik des beutetöns. *Z. Saugetierkde* **21**: 135–137.

EIBL-EIBESFELDT, I. (1963) *Ethology—The Biology of Behaviour.* New York: Holt, Rinehardt and Winston.

ELDER, W.H. (1951) The baculum as an age criterion in mink. *J. Mammal.* **32**: 43–50.

ENDERS, R.K. (1952) Reproduction in mink. *Proc. Amer. Phil. Soc.* **96**: 691–755.

ERLINGE, S. (1968) Food habits of captive otters. *Oikos* **19**: 259–270.

ERLINGE, S. (1969) Food habits of the otter (*Lutra lutra*) and mink (*Mustela vison*) in trout waters in Sweden. *Oikos* **20**: 1–7.

ERLINGE, S. (1972) Interspecific relations between otter (*Lutra lutra*) and mink (*Mustela vison*) in Sweden. *Oikos* **23**: 327–335.

ERLINGE, S. (1979) Adaptive significance of sexual dimorphism in weasels (*Mustela nivalis*). *Oikos* **33**: 233–245.

ERRINGTON, P.L. (1954) An analysis of mink predation upon muskrats in north-central United States. *Iowa State Coll. Agric. Exp. Stn. Res. Bull.* **320**: 798–924.

ERRINGTON, P.L. (1961) *Muskrats and Marsh Management.* Lincoln: University of Nebraska Press.

FAIRLEY, J.S. (1980) Observations on a collection of feral Irish mink (*Mustela vison* Schreber). *Proc. Roy. Irish Acad.* **80B**: 81–92.

FAIRLEY, J.S. & WARD, D.P. (1987) Correction factors and mink faeces. *Irish Nat. J.* **22**: 33–46.

FROST, W.E. & BROWN, M.E. (1967) *The Trout.* London: Collins.

GERELL, R. (1967a) Dispersal and acclimatisation of the mink (*Mustela vison* Schreber) in Sweden. *Viltrevy* **4**: 1–38.

GERELL, R. (1967b) Food selection in relation to habitat in mink (*Mustela vison*). *Oikos* **18**: 233–246.

GERELL, R. (1968) Food habits of the mink (*Mustela vison*) in Sweden. *Viltrevy* **5**: 119–194.

GERELL, R. (1969) Activity patterns of mink (*Mustela vison*) in southern Sweden. *Oikos* **20**: 451–460.

GERELL, R. (1970) Home ranges and movements of mink (*Mustela vison*) in Sweden. *Oikos* **21**: 160–173.

GERELL, R. (1971) Population studies of mink (*Mustela vison*) in southern Sweden. *Oikos* **8**: 83–109.

GERELL, R. (1985) Habitat selection and nest predation in a common eider population in southern Sweden. *Ornis Scand.* **16**: 129–139.

GESNER, C. (1620) *Historiae Animalium de Quadripedibus Viviparus, I.* Frankfurt. (Quoted in Youngman, 1982).

GEWALT, W.V. (1959) Beiträge zur kenntnis des optischen Differenzierungsvermögens einiger Musteliden mit besonderer Berücksichtigung des Farbensehens. *Zool. Beitr.* **5**: 117–175.

GILBERT, F.F. (1969) Analysis of the basic vocalisations of the ranch mink. *J. Mammal.* **50**: 625–627.

GILBERT, F.F. & BAILEY, E.D. (1969) Visual isolation and stress in female ranch mink particularly during the reproductive season. *Can. J. Zool.* **47**: 209–212.
GILBERT, F.F. & GOFTON, N. (1982a) Heart rate values for beaver, mink and muskrat. *Comp. Biochem. Physiol.* **73A**: 249–251.
GILBERT, F.F. & GOFTON, N. (1982b) Terminal dives in mink, muskrat and beaver. *Physiol. Behav.* **28**: 835–840.
GITTLEMAN, J.L. & HARVEY, P.H. (1982) Carnivore home range size, metabolic needs and ecology. *Behav. Ecol. Sociobiol.* **10**: 57–64.
GOODPASTER, W. & HOFFMEISTER, D.F. (1950) Bats as prey for mink in a Kentucky cave. *J. Mammal.* **31**: 319–330.
GOODWIN, G.G. (1935) Mammals of Conneticut. *Conn. Geol. Nat. Hist. Surv.* **53**: 70–71.
GOSLING, M.L. (1986) Economic consequences of scent marking in mammalian territoriality. In *Chemical Signals in Vertebrates*, D. Duvall, D. Müller-Schwarze & R.M. Silverstein (eds): New York: Plenum.
GOSSOW, H. (1970) Vergleichende Verhaltensstudien an Marderartigen. I. Uber LautäuBerungen und zum Beuteverhalten. *Z. Tierpsychol.* **27**: 405–480.
GREEN, J. (1977) Sensory perception in hunting otters *Lutra lutra* L. *Otter Trust J.* 13–16.
GREEN, J., GREEN, R. & LILES, G. (1986) Interspecific use of resting places by mink *Mustela vison* and otter *Lutra lutra*. *Vincent Wildl. Trust Rep.* **1985**: 20–26.
GREGORY, J.G. (1987) Nutritional aspects of sexual dimorphism in the American mink *Mustela vison* Schreber. Ph.D. Thesis. Durham University.
GRIGOR'EV, N.D. & EGOROV, Y.E. (1967) [The biocoenotic connections of the mink with the common otter in the Bashkirian SSR] *Sb. Trud. nauchno-issled. Inst Zhivotn. Syr'ya Pushining* **22**: 26–32.
HALL, E.R. (1982) *The Mammals of North America. II.* New York: Wiley.
HAMILTON, W.D. (1959) Food of mink in New York. *New York Fish Game* **6**: 77–85.
HANSSON, A. (1947) The physiology of reproduction in mink (*Mustela vison* Schreber) with special reference to delayed implantation. *Acta Zool.* **28**: 1–136.
HARRISON, M.D.K. & SYMES, R.G. (1989) Economic damage by feral American mink (*Mustela vison*) in England and Wales. In *Mammals as Pests*, pp 242–251, R. Putman (ed.): London: Chapman and Hall.
HATLER, D.F. (1976) The coastal mink of Vancouver Island, British Columbia. Ph.D. Thesis. University of British Columbia, Canada.
HEER, E. (1965) *Beiträge zur Säugetierkunde Süd-Bessarabiens und der Nord-Dobrudsha*. Heimatumuseum der Deutschen aus Bessarabien, Schriftenreihe D, No. 1.
HEFNER, R.S. & HEFNER, H.E. (1985) Hearing in mammals: the least weasel. *J. Mammal* **66**: 745–755.
HEIDEMANN, G. (1983) Über das Vorkommen des Farmnerzes (*Mustela vison* f. *dom.*) in Schleswig-Holstein. *Z. Jagdwissenschaft* **29**: 120–122.
HERTER, K. & KLAUNIG, J. (1956) Untersuchungen an der Retina amerikanischer Nerz (*Mustela lutreola vison* Schreb.). *Zool. Beit. N.F.* **2**: 127–143.

HEWSON, R. (1971) Some aspects of the biology of feral mink *Mustela vison* Schreber in Banffshire. *Glasgow Nat.* **18**: 539–546.
HILL, G.R. (1964) Wild mink in west Wales. *Nature in Wales,* Summer 1964, 17–18.
IRELAND, M.C. (1990) The behaviour and ecology of the American mink (*Mustela vison* Schreber) in a coastal habitat. Ph.D. Thesis. Durham University.
IVERSEN, J.A. (1972) Basal energy metabolism in mustelids. *J. Comp. Physiol.* **81**: 341–344.
JENKINS, D. & HARPER, R.J. (1980) Ecology of otters in northern Scotland. 2. Analysis of otter *Lutra lutra* and mink *Mustela vison* faeces from Deeside, N.E. Scotland in 1977–78. *J. Anim. Ecol.* **49**: 737–754.
JEWELL, P.A. (1960) The concept of home range in mammals. *Symp. Zool. Soc. Lond.* **18**: 85–109.
JOERGENSEN, G. (1985) (ed.) *Mink Production.* Hillerod, Denmark: Scientifur.
JONGE, G., CARLSTEAD, K. & WIEPKEMA, P.R. (1986) *The Welfare of Ranch Mink.* Spelderholt. Centre for Poultry Res. and Ext. Het. In Dutch.
KÖNIG, C. (1973) *Mammals.* London: Collins.
KORSCHGEN, L.G. (1958) December food habits of mink in Missouri. *J. Mammal.* **39**: 521–527.
KRAMER, D.L. (1988) The behavioural ecology of air-breathing by aquatic mammals. *Can. J. Zool.* **66**: 89–94.
KRUUK, H. & HEWSON, R. (1978) Spacing and foraging of otters (*Lutra lutra*) in a marine habitat. *J. Zool. Lond.* **185**: 205–212.
KURTÉN, B. (1968) *Pleistocene mammals of Europe.* London: Weidenfeld & Nicolson.
LAFONTAINE, L. (1988) Un nouveau venu sur le littoral: le vison d'Amérique. *Penn-ar-Bed* **125**: 77–82.
LASHLEY, K.S. (1930) The mechanism of vision. III The comparative visual acuity of pigmented and albino rats. *J. Genet. Psychol.* **37**: 481–484.
LEICHLEITNER, R.R. (1954) Age criteria in mink *Mustela vison. J. Mammal.* **35**: 496–503.
LEVER, C. (1977) *Naturalised Animals of the British Isles.* London: Hutchinson.
LEVER, C. (1978) The not so innocuous mink. *New Scientist* **78**: 481–484.
LEVER, C. (1985) *Naturalised Animals of the World.* London: Longman.
LINN, I.J. & BIRKS, J.D.S. (1981) Observations on the home range of feral American mink (*Mustela vison*) in Devon, England, as revealed by radiotracking. In Chapman J.A. & Pursley D, (eds) *Proceedings of the Worldwide Furbearer Conference,* Vol. II: 1088–1102 Frostburg, Maryland, USA.
LINN, I.J. & BIRKS, J.D.S. (1989) Mink (Mammalia; Carnivora; Mustelidae): correction of a widely quoted error. *Mamm. Rev.* **19**: 175–179.
LINN, I.J. & CHANIN, P.R.F. (1978a) Are mink really pests in Britain? *New Scientist* **77**: 560–562.
LINN, I.J. & CHANIN, P.R.F. (1978b) More on the mink menace. *New Scientist* **79**: 38–40.

LINN, I.J. & STEVENSON, J.H.F. (1980) Feral mink in Devon. *Nature in Devon* **1**: 7–27.

LOCKIE, J.D. (1959) The estimation of food of foxes. *J. Wildl. Mgmt* **21**: 224–227.

LONG, C.A. & HOWARD, G. (1987) Intraspecific overt fighting in the wild mink. *Rept. Fauna Flora Wisconsin (Uni. Wisc. Mus. Nat Hist.)* **11**: 4–5.

McCABE, R.A. (1949) Notes on live-trapping mink. *J. Mammal.* **30**: 416–423.

MACDONALD, D.W. (1978) Radio-tracking: some applications and limitations. In *Recognition Marking of Animals in Research*, B. Stonehouse (ed.): London: Macmillan.

MACDONALD, D.W. (1985) The carnivores: order Carnivora. In *Social Odours in Mammals*, Vol. 2, pp 195–205 R.E. Brown & D.W. Macdonald (eds): Oxford: Clarendon Press.

MACLENNAN, R.R. & BAILEY, E.D. (1969) Seasonal changes in aggression, hunger and curiosity in ranch mink. *Can. J. Zool.* **47**: 1395–1404.

McNAB, B.K. (1963) Bioenergetics and the determination of home range size. *Amer. Nat.* **97**: 133–140.

MARAN, T. (1990) Conservation of the European mink in Estonia. International Union for the Conservation of Nature, *Must. Viv. News* **2**, 12.

MARSHALL, W.H. (1936) Study of the winter activities of the mink. *J. Mammal.* **17**: 382–392.

MASON, C.F. & MACDONALD, S.M. (1983) *Otters—Ecology and Conservation*. Cambridge: Cambridge University Press.

MECH, L.D. (1965) Mink: the weasel of the waterways. *Anim. King.* **68**: 87–90.

MELQUIST, W.E. & HORNOCKER, M.G. (1982) Methods and techniques for studying and censusing river otter populations. *For. Wldl. Range Exp. Sta. Tech. Rep. No. 8:* 1–17.

MELQUIST, W.E., WHITMAN, J.S. & HORNOCKER, M.G. (1981) Resource partitioning and co-existence of sympatric mink and river otter populations. In *Proceedings of the Worldwide Furbearer Conference*, Chapman J.A. & Pursley P, (eds): vol. I Frostburg, Maryland, USA.

MITCHELL, J.L. (1958) Mink population study. *Montana Fish Game Proj. Rep. W49-R-7.*

MITCHELL, J.L. (1961) Mink movements on a Montana river. *J. Wildl. Mgmt* **25**: 49–54.

MOORS, P.J. (1977) Studies of the metabolism, food consumption and assimilation efficiency in the weasel *Mustela nivalis*. *Oecologia* **27**: 185–202.

MOORS, P.J. (1980) Sexual dimorphism in the body size of mustelids: the roles of food habits and breeding systems. *Oikos* **34**: 147–158.

MORRIS, P.A. (1972) A review of mammalian age determination methods. *Mamm. Rev.* **2**: 69–104.

MÜLLER, D. (1930) Sinnesphysiologische und psychologische untersuchungen an Musteliden. *Z. Vgl. Physiol.* **12**: 293–328.

MYRCHA, A. (1986) Calorific value and chemical composition of the body of the European hare (*Lepus europaeus*). *Acta Theriol.* **13**: 65–70.

NAYLOR, E. (1962) Seasonal changes in a population of *Carcinus maenas* (L.) in the littoral zone. *J. Anim. Ecol.* **31**: 601–609.

NEUMANN, D. & SCHMIDT, H.D. (1959) Optische differenzierungsleistungen von Musteliden versuche on Frettchen un Iltisfrettchen. *Z. Vgl. Physiol.* **42**: 199–205.
NILSON, G. (1980) *Facts about Furs*. Washington: Animal Welfare Institute.
NOVIKOV, G.A. (1956) *Carnivorous Mammals of the USSR*. Translation 1962, Israel Programme for Scientific Translations, Jerusalem.
OLROG, C.C. & LUCERO, M.M. (1981) *Guià de los Mamiferos Argentinos*. Argentina: Ministerio de Cultura y Educacion Fundacion Miguel Lillio Stucuman.
ORING, L.W., LANK, D.B. & MAXSON, S.J. (1983) Population studies of the polyandrous spotted sandpiper. *Auk* **100**: 272–285.
O'SHEA, T.J., KAISER, T.E., ASKINS, G.R. & CHAPMAN, J.A. (1981) Polychlorinated biphenyls in mink. In *Proceedings of the Worldwide Furbearer Conference*, Vol. I, Chapman J.A. & Pursley P. (eds): Frostburg, Maryland, USA.
PASCAL, M. & DELATTRE, P. (1981) Comparaison de différentes méthodes de détermination de l'àge individuel chez le vison (*Mustela vison* Schreber). *Can. J. Zool.* **59**: 202–211.
PEARSON, W.E. (1971) Mink. In *The UFAW Handbook on the Care and Management of Farm Animals* pp 229–238: London: Churchill Livingstone.
PETERS, R.H. (1983) *The Ecological Implications of Body Size*. Cambridge: Cambridge University Press.
PETRIDES, G.A. (1950) The determination of age and sex ratios in fur animals. *Amer. Nat.* **43**: 355–382.
PLATANOW, N.S. & KARSTED, L.H. (1973) Dietary effects of polychlorinated biphenyls on mink. *Can. J. Comp. Med.* **37**: 391–400.
POOLE, T.B. & DUNSTONE, N. (1976) Underwater predatory behaviour of the American mink (*Mustela vison*). *J. Zool. Lond.* **178**: 395–412.
POWELL, R.A. (1979) Mustelid spacing patterns—variations on a theme by *Mustela*. *Z. Tierpsychol.* **50**: 153–165.
POWELL, R.A. & ZIELINSKI, W.J. (1989) Mink respond to ultrasound in the range emitted by prey. *J. Mammal.* **70**: 637–638.
RACEY, G.D. & EULER, D.L. (1983) Changes in mink habitat and food selection as influenced by cottage development in central Ontario. *J. Appl. Ecol.* **20**: 387–402.
RALLS, K. & HARVEY, P.H. (1985) Geographic variation in size and sexual dimorphism of the North American weasel. *Biol. J. Linn. Soc.* **25**: 119–167.
RITCEY, R.W. & EDWARDS, R.Y. (1956) Live trapping mink in British Columbia. *J. Mammal.* **37**: 114–116.
ROBINSON, I.H. (1987) Olfactory communication and social behaviour in the mink (*Mustela vison*). Ph.D. Thesis. University of Aberdeen.
ROMANOWSKI, J. (1990) Minks in Poland. *Must. Viv. News.* **2**: 13.
ROSENWEIG, M.L. (1966) Community structure in sympatric carnivores. *J. Mammal.* **47**: 602–612.
ROSENWEIG, M.L. (1968) Strategy of body size in mammalian carnivores. *Amer. Midl. Nat.* **80**: 299–315.

RUPRECHT, A.L., BUCHALCZYK, T. & WOJCIK, J.M. (1983) Wystepowanie norek (Mammalia: Mustelidae) w Polsce. [The occurrence of minks in Poland.] *Przeglad Zoologiczny* **27**: 87–99.

RUST, C.C., SHACKLEFORD, R.M. & MEYER, R.K. (1965) Hormonal control of pelage cycles in mink. *J. Mammal.* **46**: 549–565.

SARGEANT, A.B., SWANSON, G.A. & DOTY, H.A. (1973) Selective predation by mink on waterfowl. *Amer. Midl. Nat.* **89**: 208–214.

SCHREIBER, A., WIRTH, R., RIFFEL, M. & ROMPAEY, VON H. (1989) *Weasels, Civets, Mongooses and Their Relatives: An Action Plan for the Conservation of Mustelids and Viverrids*. Switzerland: International Union for Nature Conservation.

SCHUSTERMAN, R.J. & BALLIET, R.F. (1970a) Visual acuity of the harbour seal and Stellar sea lion underwater. *Nature Lond.* **226**: 563–564.

SCHUSTERMAN, R.J. & BALLIET, R.F. (1970b) Conditioned vocalisation as a technique for determining visual acuity thresholds in sea lions. *Science, NY* **169**: 498–501.

SCHUSTERMAN, R.J. & BALLIET, R.F. (1971) Aerial and underwater visual acuity in the Californian sea lion *Zalophus californianus* as a function of luminance. *Ann. NY Acad. Sci.* **188**: 37–46.

SEALANDER, J.A. (1943) Winter food habits of mink in southern Michigan. *J. Wildl. Mgmt* **7**: 411–417.

SHACKLEFORD, R.M. (1950) *Genetics of the Ranch Mink*. Pilsing.

SHACKLEFORD, R.M. (1952) Superfoetation in the ranch mink. *Am. Nat.* **86**: 311–319.

SHUBIN, I.G. & SHUBIN, H.G. (1975) Sexual dimorphism and peculiarities in mustelines. *Z. Obshcei Biol.* **36**: 283–290.

SIDWELL, V.D. (1974) Composition of the edible portion of crustaceans, fin fish, and molluscs. *Mar. Fish. Rev.* **36**: 21–35.

SINCLAIR, D.G., EVANS, E.V. & SIBBALD, I.R. (1962) Influence of weight digestible energy and apparent digestible nitrogen in diet on weight gain, feed consumption and nitrogen retention of growing mink. *Can. J. Biochem. Physiol.* **40**: 1375–1389.

SINCLAIR, W., DUNSTONE, N. & POOLE, T.B. (1974) Aerial and underwater visual acuity in the mink *Mustela vison* Schreber. *Anim. Behav.* **22**: 965–974.

SKIRNISSON, K. (1980) The mink in Iceland, 'Villt Spendyr', *Rit Landverndar* **7**: 80–94.

SMITH, K.U. (1936) Visual discrimination in the cat. IV The visual acuity of the cat in relation to stimulus distance. *J. Genet. Psychol.* **49**: 297–313.

SPONG, P. & WHITE, D. (1971) Visual acuity and discrimination learning in the dolphin (*Lagenorhynchus obliquidens*). *Exp. Neurol.* **31**: 431–436.

STEPHENSON, R., BUTLER, P.J., DUNSTONE, N. & WOAKES, A.J. (1988) Heart rate and gas exchange in freely diving American mink *Mustela vison*. *J. Exp. Biol.* **134**: 435–442.

SUNDQVIST, C. & GUSTAFSSON, M. (1983) Sperm test—a useful tool in breeding work of mink. *J. Sci. Agric. Soc. Finland* **55**: 119–131.

SUNDQVIST, C., LE GRANDE, C.E. & BARTKE, A. (1988) Reproductive endocrinology of the mink (*Mustela vison*). *Endocr. Rev.* **9**: 247–266.

TATE, G.H.H. (1931) Random observations on the habits of South American mammals. *J. Mammal.* **12**: 248–256.

TERNOVSKY, D.V. (1977) *Biologia kuniceobraznyh (Mustelidae)*. Novosibirsk. Izd: Nauka.

THOMPSON, H.V. (1962). Wild mink in Britain. *New Scientist* **13**: 130–132.

TUMANOV, I.L. & ZVEREV, E.L. (1986) Sowremennoje rasprostranenije i tschislennost jewropejskoj norki (*Mustela lutreola*) w. USSR. *Zoologitscheskij Shurnal* **65**: 427–435.

VAN BREE, P.H.J. & SAINT-GIRONS, M.C. (1966) Donnees sur la repartition et la taxonomie *Mustela lutreola* (Linnaeus 1761) en France. *Mammalia* **30**: 270–291.

VAN DEN BRINCK, F.H. (1967) *A Field Guide to the Mammals of Britain and Europe*. London.

VENGE, O. (1971) Reproduction in the mink. *Kgl. Vet-og Landbohojst Arsskr.* **1973**: 95–146.

WALLS, G.L. (1965) *The Vertebrate Eye and its Adaptive Radiation*. New York: Cranbrook Institute of Science.

WALTON, K.C. (1968) Studies on the biology of the polecat *Putorius putorius*. M.Sc. Thesis. University of Durham.

WARD, D.P., SMAL, C.M. & FAIRLEY, J.S. (1986) The food of mink *Mustela vison* in the Irish Midlands. *Proc. Roy. Irish Acad.* **86B**: 169–182.

WATERS, J.H. & RAY, C.E. (1961) Former range of the sea mink. *J. Mammal.* **42**: 380–381.

WEST, N.H. & VAN VLIET, B.N. (1986) Factors influencing the onset and maintenance of bradycardia in mink. *Physiol. Zool.* **59**: 451–463.

WHITE, D., CAMERON, N., SPONG, P. & BRADFORD, J. (1971) Visual acuity of the killer whale *Orcinus orca*. *Exp. Neurol.* **32**: 230–236.

WHITMAN, J.S. (1981) Ecology of the mink (*Mustela vison*) in west-central Idaho. M.Sc. Thesis. University of Idaho, USA.

WIIG, O. (1982) Sexual dimorphism in the skulls of feral mink. *Zool. Scripta* **11**: 315–316.

WIIG, O. (1986) Sexual dimorphism in the skull of mink (*Mustela vison*), badgers (*Meles meles*) and otters (*Lutra lutra*). *Biol. J. Linn. Soc.* **87**: 163–179.

WILLIAMS, T.M. (1983a) Locomotion in the mink—a semi-aquatic mammal. I. Swimming energetics and body drag. *J. Exp. Biol.* **103**: 155–168.

WILLIAMS, T.M. (1983b) Locomotion in the mink—a semi-aquatic mammal. II. The effect of an elongate body on running energetics and gait patterns. *J. Exp. Biol.* **105**: 283–295.

WILSON, D.S. (1975) The adequacy of body size as a niche difference. *Amer. Nat.* **109**: 769–784.

WILSON, K.A. (1954) The role of mink and otter as muskrat predators in northeastern North Carolina. *J. Wildl. Mgmt* **18**: 199–207.

WISE, M.H., LINN, I.J. & KENNEDY, C.R. (1981) Comparison of feeding ecology of the mink (*Mustela vison*) and the otter (*Lutra lutra*). *J. Zool. Lond.* **195**: 181–213.

WOEBESER, G., NIELSEN, N.C. & SCHIEFER, B. (1976) Mercury and mink. I. The use of mercury contaminated fish as a food for ranch mink. *Can. J. Comp. Med.* **40**: 30–33.

WREN, C.D., STOKES, P.M. & FISCHER, K.L. (1986) Mercury levels in Ontario mink and otter relative to food levels and environmental acidification. *Can. J. Zool.* **64**: 2854–2859.

WÜSTEHUBE, C.V. (1960) Beiträge zur kenntnis besonders des spiel-und beuteverhaltens einheimischer Musteliden. *Z. Tierpsychol.* **17**: 579–613.

YEAGER, L.E. (1943) Storing of muskrats and other food by mink. *J. Mammal.* **24**: 100–101.

YOUNGMAN, P.M. (1982) Distribution and systematics of the European mink *Mustela lutreola* Linnaeus 1761. *Acta Zool. Fenn.* **166**: 1–48.

Species Index

Abramis brama 85
Actitis macularia 192
Agelaius phoeniceus 83
Alca torda 77
Alnus glutinosa 125
Amazonian weasel 13, 18, 19
Amblonyx cineria cineria 51
Ambystoma maculatus 86
Ambystoma tigrinum 75
American wigeon 76
Amphibia 78–9, 86
Anas acuta 83
Anas clypeata 76
Anas crecca 76
Anas discors 74
Anas stepera 76
Anseriformes 92
Anthus 75, 77
Apodemus sylvaticus 77
Arctic tern 77, 192
Ardea cinerea 195
Arvicola terrestris 193
Astacus astacus 87
Aythya affinis 76
Aythya americana 76
Aythya valisineria 76

Baboon 60
Badger 18, 59, 62, 88, 106, 124, 140, 194
Bank vole 81
Bat 77
Baylisascaris devosti 162
Bear 18
Beaver 166
Biting lice 160
Black footed ferret 19

Black guillemot 191, 192
Black-headed gull 77
Blacknose dace 85
Blenny 86
Blue-winged teal 74, 75, 76, 83
Bluebird 83
Bombus jounelis 88
Bream 85
Brook trout 84
Brown hare 71, 77, 90–91, 98
Brown rat 159
Bullfrog 87
Bullhead 85, 162
Bumblebee 88
Burbot 75, 85
butterfish 86

California gull 75
California sea lion 51
Cambarus 87
Cancer productus 87
Canis lupus 162
Canvasback 76
Carcinus maenas 87, 90
Cardinal 83
Cardinalis 83
Carp 184
Castor fiber 166
Cat 46, 47, 48, 202
Catostomus 84
Catostomus macrocheilus 197
Cephus grylle 191, 192
Cestoda 162
Charadriiformes 83
Chinese weasel 21
Chrysemys 87
Ciliata mustela 85

217

Cladium jamaicense 101
Clawless otter 51
Clethrionomys gapperi 77
Clethrionomys glareolus 81
Colombian weasel 13
Coloptes auratus 83
Columba 77
Common gull 192
Common pumpkinseed 85
Common shiner 85
Common tern 7, 192
Coot 74, 75, 82, 83, 92, 190, 191
Corvus 75, 77
Cotton rat 74
Cottontail rabbit 77, 93
Cottus bairdi 197
Cottus gobio 85
Coypu 187
Crabs 16, 72–3, 87, 90–1, 114
Crayfish 16, 87, 90, 102
Crow 75
Crustacea 16, 76, 87, 90–1, 92, 94, 102, 108
Ctenophthalmus nobilus vulgaris 159, 160
Cyclopterus lumpus 86

Dace 38
Deer mouse 77
Diotophyma renale 162
Dog 18, 202, 204
Dolphin 50, 51
Domestic fowl 98
Domestic stock 193–5
Dormouse 82
Ducks 74–6, 82–4, 92, 98, 190–1
Dytiscus 87

Eagle owl 162
Eel 63, 64, 85, 86, 90, 98, 190, 197
Eider duck 192
Erinaceus europeas 77
Esox lucius 85
Eumetopias 51
European desman 192
European mink 2, 6, 11–12, 19, 21, 29–32, 39, 93, 101, 115, 131, 149, 196
Eurycea 86
Everglades mink 22

Ferret 14, 34, 47, 52
Field vole 82, 98, 108

Fisher 166
Fleas 159, 160
Flicker 83
Fox 88, 194, 202
Freshwater fish 84–5, 190
Frog 74, 75, 78–9, 86, 93
Fulica americana 83
Fundulus heteroclitus 84

Gadwall 75, 76
Galemys pyrenacius 192
Gallus domesticus 98
Garter snake 87
Gasterosteus aculeatus 85
Giant otter 1
Golden shiner 85
Grayling 85
Grebe 75
Green-winged teal 76
Grey partridge 75, 77
Ground squirrel 77, 92
Guillemot 77
Gulo gulo 18, 162

Haematopus oestralegus 77
Haliotis kamschatkana 88
Harbour seal 51
Hare, brown 62, 71, 77, 90–1
Harvest mouse 82
Hedgehog 77, 159
Heron 195
Herpestidae 62
Herring gull 77, 84, 97, 99, 149, 192, 202
Himalayan porcupine 187
Horned dace 85
Hungarian partridge 75
Hyla catesbeiana 86
Hyla crucifera 86
Hystrix brachyure 187

Ide 85
Invertebrates 87–8, 91
Ixodes hexagonus 159, 160
Ixodes ricinus 159, 160
Ixodes spinitalpis 160

Japanese mink 21

Killer whale 51
Kokanee 197
Kolinsky 21

Lagenorhynchus obliquidens 51
Lagomorph 62, 91, 94, 95, 97, 106, 114, 192
Lapwing 77
Large scale sculpin 197
Larus argentatus 192
Least weasel 18
Leopard frog 87
Lepomis gibbosus 84
Lepus capensis 77
Lesser scaup 76
Leuciscus idus 85
Leuciscus leuciscus 38
Lice 160
Ligia oceanica 88
Ligia pallasii 88
Lipophrys pholis 85
Lota lota 75, 85
Lumpfish 86
Lutra canadensis 14, 196
Lutra lutra 14, 20
Lutreola 20
Lutreola 20
Lutrinae 18
Lynx 162
Lynx lynx 162

Mallard 74, 75, 76
Marine fish 85–6
Marten 18, 166
Martes 18
Martes martes 18, 166
Martes pennanti 166
Martes zibellina 20
Meadow vole 77
Megabothris walkeri 160
Meles meles 20, 140
Melinae 18, 62
Mephitinae 18, 62
Mephitis mephitis 140
Micromys minutus 82
Micropterus dolomieui 84
Microtis agrestis 82, 98, 108
Microtus arvalis orcadensis 204
Microtus ochrogaster 77
Microtus pennsylvanicus 77
Minnow 68, 85
Mole 82, 93
Mongoose 62
Moorhen 82, 83, 92, 190–1
Mottled sculpin 197
Mountain weasel 19
Mummichog 85

Muscardinus avellanarious 82
Muskrat 62, 74, 80–1, 90, 92, 93, 101, 105, 187
Mustela africana 13, 14, 18, 19
Mustela altaica 19
Mustela erminea 19, 20, 45, 140, 201
Mustela eversmanni 19
Mustela felipei 13, 14
Mustela frenata 14, 19
Mustela furo 14
Mustela lutreola 2, 6, 11–12, 19, 21, 29–32, 39, 101, 140
Mustela lutreola cyclipena 29
Mustela lutreola lutreola 29
Mustela lutreola novikovi 29
Mustela lutreola transylvania 29
Mustela lutreola turovi 29
Mustela macrodon 32–3, 164
Mustela nigripes 19
Mustela nivalis 19, 20, 45, 140
Mustela nivalis rixosa 18
Mustela putorius 14, 19, 20, 140, 201
Mustela sibirica 19, 21
Mustela vison lowii 20, 22
Mustela vison aesturina 22
Mustela vison ariakensis 22
Mustela vison enurgumenas 22
Mustela vison evagor 22
Mustela vison evergladensis 22
Mustela vison ingens 166
Mustela vison lacustris 20, 22
Mustela vison latifera 20, 22
Mustela vison lutensis 22
Mustela vison melampeplus 22, 166
Mustela vison mink 22
Mustela vison nesolestes 22
Mustela vison vison 20, 22, 166
Mustela vison vulvivaga 22
Mustelidae 4, 18, 68, 124
Mustelinae 18, 19, 62
Myocastor coypus 187
Myotis 77
Myxocephalus scorpio 86

Nematoda 162
Neuroptera 87
Northern kelp crab 87
Northern shoveller 76
Nosopsyllus fasciatus 159, 160
Notemigonus crysoleucas 84
Notropis cornutus 84

Oncorhynchus nerka 197

Ondatra zibethicus 187
Opheodrys vernalis 86
Orca orca 51
Orchoppeas 160
Orkney vole 204
Oryctolagus cuniculus 75, 76
Otter 1, 5, 10, 14, 15, 16, 18, 25, 38, 45, 50, 51, 56, 59, 64, 68–69, 124, 166, 187, 190, 196–201, 202
otterhounds 200, 201, 203
Ovis aries 77
Oystercatcher 77

Painted turtle 87
Paleopsylla minor 160
Panda 18
Partridge 193
Passerines 83, 98
Perca fluviatilis 63, 85
Perch 63, 85
Perdix perdix 75, 77, 193
Peromyscus 77
Phasianus colchicus 75, 77, 193
Pheasant 75, 77, 193, 195
Phoca vitulina 51
Pholis gunnelus 86
Phoxinus phoxinus 68, 85
Pied-billed grebe 83
Pike 85
Pine marten 124
Pintail 76, 83
Pipit 75
Plecoptera 87
Plesictis 1
Podilymbus podiceps 83
Polecat 1, 16, 36, 46, 47, 49, 62, 201
Porpoise 50
Prarie vole 77
Puffin 192
Pugettia producta 87

Rabbit 19, 62, 63, 64, 71, 75, 77, 80, 81, 88, 90–1, 93, 94, 95, 97, 98, 101, 102, 105, 108, 114, 116, 121, 125, 131, 132, 136, 187, 188, 192, 197, 198, 201
Ralliformes 92
Rana arvalis 87
Rana pipiens 87
Rana temporaria 87
Rat 47, 63, 71, 77, 116, 197, 204
Rattus norvegicus 77

Rattus rattus 82
Razorbill 77
Red backed vole 77
Red bellied snake 87
Red crab 87
Red deer 60
Red-winged blackbird 75, 83
Redhead 76
Redsided shiner 196
Reptiles 86–7
Rhesus monkey 47
Rhinichthys atratus 84
Rhinichthys osculus 197
Richardsonius balteatus 196
Roach 85
Rockling 86
Rodents 64, 80–1, 91
Roundworms 160–2
Rudd 85
Ruddy duck 75, 76, 83
Rutilus rutilus 85

Sable 20
Salix 100, 125
Salmo salar 84
Salmo trutta 84
Salmon 85, 190
Salmonids 63, 85, 86, 125, 190, 197, 195
Saterna 77
Scardinius erythrophthalmus 85
Scolopax rusticola 77
Sea lion 50
Sea mink 33
Sea otter 124
Sea scorpion 86
Sea slater 86
Sea trout 85, 190
Seabirds 75, 77, 83–4, 97, 191–2, 197
Seal 50
Semotilus atromculatus 84
Shag 192
Sheep 77
Sheep tick 159
Shore crab 87, 90
Shrew 82
Sialia 83
Siberian weasel 19, 21
Sinus worm 161
Skrjabingylus nasicola 161
Skunk 18, 62
Smallmouth bass 84
Smooth green snake 87

Sorex araneus 82, 92
Sorex minutus 82, 92
Speckled dace 197
Spermophilus 77
Spermophilus tridocemlineatus 75
Spotted sandpiper 192
Spring peeper 86
Squirrel 49, 106, 197
Starling 98
Steller sea lion 51
Steppe polecat 19
Sterna hirund 192
Sterna paradisea 84
Stickleback 85
Stoat 5, 19, 45, 46, 62, 94, 124, 140, 201
Storena 87
Striped skunk 140
Sturnus vulgaris 98
Sylvilagus floridanus 77

Talpa europea 82, 93
Tapeworms 160–2
Taurulus bubalis 85
Thamnophis 87
Thirteen-lined ground squirrel 75
Thymallus thymallus 85
Tick 159, 160
Tiger salamander 75, 87
Triachdectes 160

Trout 38, 98, 195
Typhloceras poppei 160

Uria aalge 77

Vanellus vanellus 77
Viverra lutrola 29
Vulpes vulpes 88

Water beetle 87
Water vole 82, 193
Waterfowl 74–6, 79, 92, 98, 190–1, 194
Weasel 1, 5, 45, 46, 53, 62, 64, 94, 124, 140
Whale 50
White-footed mouse 77
Willow 100
Wolf 162
Wolverine 18, 162
Wood duck 75
Woodcock 77
Woodmouse 77

Xanthocephalus xanthocephalus 83

Yellow-headed blackbird 83
Yukon mink 22

Zalophus californianus 51

Index

Acidification 162
Accommodation 50–2
Activity 57, 87, 97, 100–1
Activity patterns 59, 106–15
   tidal cycle 114–5
   of individuals 112–3
   of European mink 115
   periodicity 113
   rhythms 106–7
   seasonal effects on 106, 108, 113–4
   types of 108, 122
   weather 114
Adaptations 1, 3, 34–61, 59
   aquatic 3, 18, 50–2
   pelage 34
   respiratory 39–43
   sensory 45–52
Adaptive radiation 1, 19, 57
Age
   determination, methods of 118, 151–5
   longevity 155, 181, 184
   population structure 118, 158–9
   ratios 151
Agricola 29
Aggression 2, 11, 131, 133, 137, 158, 188
   intrasexual 59–61, 131, 135–9
   intersexual 95, 135, 141–3
   observations of 135–9
Aggressive vocalisation 53
Alaskan mink 22
Aleutian disease 174, 184
Aleutian gene 172–4
Allele 168, 169, 170, 171, 172
Aluminium contamination 162
Amazonian weasel 13

Amphibia, in the diet of 78–9, 86
Amphibious vision 50–2
Anaesthesia, for handling live mink 117–8, 121
Anal scent glands 54–5, 139
Anatomy
   anal glands 54
   baculum
   eye 48–52
   reproductive system 140–1
   skeleton 8–13
   teeth 10–11, 153–4
   skin 34–36
Annual cycle 8, 110, 124, 176–8
Appearance 1, 4
Arhythmia 40
Austria 31

Baculum
   morphology 141
   use in aging 118, 153, 155
Badgers 59, 62, 106, 140, 194
Barns 105
Bats, in the diet of 81
Behavioural disorders 183
Bering land bridge 20
Biometrics 7
Bird eggs, in the diet of 75, 98
Birds, hunting of
Birds, in the diet of 82–4, 91–9
Birth 118, 146, 147, 148, 177
   timing of, 146
Bite power 9
Blood 19, 40, 42–3, 45
Bodyweight
   European mink 8

Bodyweight – (contd)
  American
    females 7–8
    growth rates 7–8, 148–9
    males 7–8, 60
Bodyshape, evolution of 57
Botulism 184
Bounty schemes 28
Bradycardia 40–3
Breath-holding 39, 43
Breeding (see also reproduction)
  effect of isolation 182–3
  of 'mutation' mink 35, 167–75
  seasonal breeding 141, 188
Buoyancy 39–45

Caches, content of 74–7, 88
Caching of food 73–7, 82, 87, 88
Caging size 182
Canines 10–11, 59, 95, 142, 152
Carnassial teeth 10–11, 95, 152–4
Carrion, in the diet of 17, 80–1, 84, 88, 92, 94, 192
Censusing populations 156–7
Chinese weasel 21
Chromosome number 19, 168
Chromosomes 19, 168
Cine-filming 64
Classifying mink 18–22
Claws 4, 36, 71
Climbing ability 2, 78–9, 204
Coarse fish 63, 64, 84–6, 90, 190, 195
Coat 4, 29 (see also fur)
  exploitation for fur 30, 33, 80, 151, 163–6
  colouring 4, 34, 167, 169–75
  genetics of 167, 169–75
  guard hairs 34
  markings 6, 12, 115, 117
    ventral spot patterns 6, 115, 117, 118
  moult 6, 35–6
  quality 4, 22, 166, 167, 179
  summer 6
  varieties 4
  winter 7
Co-dominant allele 168–9
Co-dominant mutations 172
Colombian weasel 13
Commercial diet 176–7
Communication
  olfactory 54–5 (also see scent-marking)

  vocal 52–4
Comparison of American and European mink 2, 11, 13
Competition 2, 7, 21, 57, 58–9
  and body size 57, 58, 93–7
  between American and European mink 29–33
  for food 58–9, 124
  for mates 60–61, 139, 143, 144
  for territories 124, 131
  interspecific 59, 158, 189, 196–201
  intraspecific 1, 58, 59, 60, 93–97, 124, 134, 143, 144, 151
Complementary genes 172–4
Confinement 182
Congenital deformities 183
Conservation 22, 29–33, 185
Control operations 23, 26, 27, 28, 202–4
Copulation 135, 141–3,
Courtship 142, 143
Cranial morphology, see skull
Crustacea 16, 76, 87, 90–1, 92, 94, 102, 108

Darwin 1, 2, 184
Daylength
  effect on moult 35–6
  effect on reproduction 141, 146
DDT 162, 200
Deafness 174
Decline, of the otter 199–200
Delayed implantation 140, 145, 146, 149
Denmark 28
Dens
  breeding 16
  distribution 104–7, 125
  selection 100–1, 104
  sites 100, 101
  structure 101, 104
Den utilisation 104–7, 108, 112, 199
  inter-den movements 105–6, 132–3
Dental formula 10
Dentition 10 (also see teeth)
Diet 64–99 (see also foraging, hunting, prey)
  availability of prey 63
  commercial 176–7
  effect of location 89–93
  impact on prey populations 80–1, 82, 189–93
  individual differences 96–7

preferences 95
seasonal differences 78–9, 87, 89–99
sex differences 93–7
techniques of analysis 74–6
  faecal analysis 76
types of prey
  Amphibia 78–9, 86–7
  passerines 83, 98
  Crustacea 16, 76, 87, 90–1, 92, 94, 102, 108
  carrion 17, 80–1, 84, 88, 192
  freshwater fish 63, 64, 78–9, 84–6, 90, 190, 195
  invertebrates 88, 91
  mammals 78–9, 80–2, 91, 98
  marine fish 78–79, 91
  muskrats 62, 74, 80–1, 82, 90, 92, 93
  rabbits 62, 63, 64, 71, 75, 77, 80
  reptiles 87
  rodents 80–1, 91
  seabirds 75, 77, 83–4, 197
  waterfowl 74–6, 79, 82–4, 98, 190–1, 194
Dietary overlap with otter 196–201
Dihybrid crosses 172–3
Diseases 181, 183–4,
  Aleutian disease 174, 184
  botulism 184
  Chastek's paralysis 184
  distemper 184
  steatitis 184
  Wet Belly (urinary incontinence) 184
Dispersal 27, 61, 112, 119, 130, 133, 148, 149, 157, 188, 189, 192, 195, 204
Distribution 21–33
  American mink 21–8
    Eastern Europe 28
    England 23–5
    Finland 28
    Iceland 28
    Ireland 26
    Denmark 28
    North America 21–2
    Norway 27–8
    Russia 28
    Scotland 25–6
    Sweden 26–7
    Wales 23–5
    Western Europe 26
  European mink 29–32
    Eastern Europe 31

    Finland 30
    France 31–2
    Russia 29–30
    Spain 32
Diurnal vision 45, 47, 48–9
Diving behaviour 40–3, 45, 64–73
  bradycardia 40–3
  dive duration 43, 64, 66–73
    comparison with otter 68–9
    effect of current flow 69–71
    effect of water depth 45, 69–71
  field observations of, 64, 72, 150
  laboratory studies on, 40–3, 45, 64–71
  searching strategy 45
DNA 19
Domestication 163–7, 176, 182
Dominant genes 169
Drowning 41–2

Ear tags 6, 115, 117, 118, 157
Economic importance 1, 34, 163–6, 187, 189, 194–5, 202–3
Ectoparasites 159–60
EMBA (Mutation Mink Breeders Association) 179
Endangered status 22, 29–33
Endoparasites 160–2
Energy and metabolism 44
  and choice of prey 98
  and home range size 123
  thermoregulation 57
Energy requirements 60, 97–9, 113
Energy expenditure 44–5
Energetics
  basal metabolism 44, 57
  effect of body shape 44, 57
  cost of locomotion 44, 57
  cost of sexual dimorphism 2, 59–61
England 23–5
Escapes 2, 22, 23, 26, 27
Ethics of wearing fur 184–6
Everglades mink 22
Evolution 1
  adaptations for aquatic hunting 3, 18, 50–52
  Mustelidae 1, 18–19
  of body shape 1, 44, 57
  of sexual dimorphism 58–61
Evolutionary theory 1, 46, 57, 58–61, 124, 144

European mink 2, 39
  coat 4
  comparison with American mink 2, 11
  distinguishing from American mink 6, 11–3
  diet 93
  distribution 29
  gestation 149
  habitat preference 101
  phylogeny 19
  reproduction 149
Exploitation 30, 33, 34, 157–8
Extinction 30, 33, 164
Extinct relatives 18, 19
Eye structure 48–52

Faecal analysis 74
  problems of interpretation 76–7
  technique 76–7
Faeces 15–16
FBA (Fur Breeders Association) 179
Fecundity 147, 157
Feeding behaviour 63, 97–9 (see also diet, hunting, foraging, predation)
Feet 4, 14–5
Feral populations 22
  origin of, 21, 22–8
Fertility 144, 174
Fertilisation 143, 144, 177
Fighting (see aggression)
Finland 28, 30
Fish, in the diet 63, 64, 78–9, 84–6, 89–99, 190, 195
Fleas 159–60
Flukes 160–1
Food 62–99 (see also, commercial diet, foraging, hunting, predation)
Food remains (at dens) 74–7
Food storage, see caching
Food supply 57, 59, 61, 124
Foraging 60, 74, 102, 122, 123 (see also hunting, predation)
  and home range size 125
  timing of 108–114
Foot prints 14
Fossil remains 18, 19, 21
France 31–2
Frogs, in the diet of 74, 75, 78–9, 86–7, 93
Fur 34–6 (see also coat)
  breeding 35, 167–75
  colour 4, 34, 167–95
  density 34
  exploitation for 33, 80, 151, 163–6
  growth 35
    hormonal control of, 36
    effect of daylength on, 35–6
  harvest
    American mink 34, 157–8
    European mink 30
  marketing 179–80
  moult 6, 35–6, 177
  processing 179
  production 179–80
  ranching 166–86
    decline 25, 27
    origin of, 21, 22, 166–7
  structure 34–6
    guard hairs 34
    underfur 34
  trapping for, 163–6
  trade 35, 163–86
    history of, 163–7
Furring-up 177

Gape 10
Genes 4, 57, 145, 168, 169, 170, 172, 173, 174, 181
  co-dominant mutations 172
  dominant mutations 169
  recessive mutations 169
Genetics 167–75
  of coat colour 167, 169–75
Genotypes 169–75
Gestation 60, 99, 140, 145–6, 149
GLAMA (Great Lakes Mink Association) 179
Glands 15
  anal 54–55, 139
  cutaneous 54, 55
  pituitary 36, 141, 143
  proctodeal 15
  prostate 140
Grooming 122
Growth
  fur 35–6
  of kits 148–9
  rate 7–8, 57, 61
  sex differences 7–8, 61, 148–9

Habitat
  destruction of, 102
  coastal 100–3

human disturbance of, 102, 125
improvement, of 100, 102
lacustrine 100–2
marsh 80–81, 101
requirements 100–6
riverine 100–2
utilisation, of 100–3
vegetative cover 100–2, 125, 132
water quality 100–1
    eutrophic 27, 84, 100–1, 125, 126, 197
    oligotrophic 84, 100–1, 125, 190, 198
Haemoglobin 40, 43
Handling techniques 117–8
Head 1, 4, 10, 48, 95
Head-dipping 43, 44, 64, 66
Hearing
    deafness 174
    sensitivity 53–4
    vocal communication 52–3
Heart rate 40–4
Heavy metal contamination 162
Heredity 167 (see genetics)
Home ranges 123–33
(see also territorial behaviour)
    and body size 123
    core areas 104, 131, 132
    defence of 124, 134–9
    definition 123
    determination of, 115–23
    effect of habitat type 124–8
        coastal 126–8
        lacustrine 126–7
        riverine 125–7
    European mink 131
    energetics, consideration of 123
    intensity of use 131
    intersexual variation 124, 128, 135
    movements within, 131–3
    residency 128
    seasonal changes in 128
    size 125–8, 134
    scent marking 54–55
    transient individuals 130, 137
Hormones 36, 141, 143, 146, 152
    oestrogen 141, 146
    progesterone 146
    testosterone 153
Hudsons' Bay Company 164, 179
Human attitudes 2, 187–9
Hunting 27, 28, 30, 83, 200, 203–4
    (also see persecution)

Hunting behaviour 83 (also see foraging, predation)
    aquatic 63–71
    caching 73–7, 82, 87, 88
    capture of prey 62–3
    effects on prey populations
        fish 190, 195
        birds 62–3
        mammals 80–1
    efficiency 63, 68
    feeding 63, 97–9
    killing bite 17, 62
    observations in wild 64, 71–3, 94, 101
    senses used during, 45–6, 54, 55–6, 73
    sets 59
    strategy 37, 62–73
    success 66–73
    surplus killing 73–5, 193, 202
Hypermetropia (longsightedness) 50

Ice Age 20
Iceland 28
Immunology 19, 20
Implantation 140, 144, 145, 146, 149, 158
Individual recognition 6, 139
Induced ovulation 143, 144
Innoculation 184
Instinctive behaviour 62
Intersexual interactions 95, 134, 135, 139, 141–5
Intrasexual interactions 134, 135–9
Invertebrates, in the diet 87, 91
Ireland 26
Islands 26, 28, 30, 87, 97, 99, 126, 136, 142, 146 149, 191, 192, 202, 204
Isolation 182–3

Japanese mink 21
Jaws 4, 10–11
Jaw musculature 10, 12, 95
Juveniles 7, 115, 118, 119, 125, 130,133, 136, 143, 151, 153, 154, 155, 157, 161, 195

K-selection 57
Karyotypes 19, 168
Killing bite 17, 62, 63
Kit growth 148–9
Kolinsky 21

Lactation 60, 118, 128, 133, 147, 148
Lagomorphs 62, 94, 106, 114
Learning ability 62, 67–8
Leg-hold traps 185
Legislation 23, 205
Length 4
Lens 49, 50–2
　use in aging 155
Lice 160
Licensing 23, 26, 187, 205
Lifestyle 100–39
Limbs 1, 4, 19, 36–9, 45, 62, 185
Lip patch 6, 12
Litter frequency 141, 188
Litter size 57, 167
　in captivity 147, 158
　in wild 118, 147–8, 149
Live-trapping 115–20, 155–156
Locomotion 36–9
　aquatic 37–9
　　buoyancy 39, 45
　　drag 45
　　posture 37
　　gaits
　　　surface swimming 37
　　　underwater 37–9
　　speed 38–9
　terrestrial 36–8
　　gaits 36
　　limb movements 36–7
　　posture 36
　　speeds 36, 101, 132
Longevity 155, 184
Long-sightedness 50
Lutrinae 18
Lynx 186

Mammals, in the diet of 74–5, 76–9, 80–2, 91, 98, 192–3
Marking 157
　ear tags 6, 115, 117, 118, 157
Mating 174
　behaviour 139, 142, 141, 144
　European mink 149
　season 105, 119, 128, 155, 188,
　system 60, 144, 146
　timing of 141, 144
　regime on ranches 175, 176–8
　wounds 118, 142
Measurements 4, 7
Melinae 18
Mendel 167
Mendelian inheritance 167–9

Mephitinae 18
Mercury contamination 162
Metabolic rate 44, 57
Middens 87 (see also caching)
Milk 148
Mink (Importation & Keeping Order) 1932 23, 205
Mites 159–60
Mortality
　adult 158–9, 184
　juveniles 159
　kits 147, 156, 184
Moult 6, 35–6, 177
　environmental control 35–6
Movement detection 45–6
Movements (see also home-range and territory)
　dispersal 27, 61, 112, 119, 130, 133, 148, 149, 157, 188, 189, 192, 204
　downstream 14
　methods of study 115–8
　patterns 100–3, 132
　speed 101, 132
Muscles
　arrector-pili 34
　digastric 10, 12
　temporalis 10
　zygomatico-mandibularis 10
Muskrat 62, 74, 80–1, 82, 90, 92–3, 101, 187
Mustelidae 4, 18, 68, 124
Mustelinae 18, 19, 62
'Mutation' mink breeding 167
Muzzle 56

Nocturnal vision 45, 47, 48–9
Norway 27–8
North America 21–2
Nutrition, (see also feeding, diet) 176–7

Odour 15–6, 54, 116
Oestrus 141, 143, 149
Olfaction 54–5
Olfactory
　communication 54–5
Ontogeny 62, 64
Origin of feral populations 21, 22–8
Os-penis (see baculum)
Otter 14, 16, 18, 25, 59, 64, 68–9, 106, 124, 196–201
Otterhounds 200, 201, 203
Ovulation 143, 144, 146

Oxygen conservation 40
  constraint 66–7
  storage 39, 42, 43, 66, 71
  uptake 42, 44–5

Parasites 152–62
  ectoparasites 159–60
  endoparasites 160–2
Parental care 57
Parturition 118, 146, 148, 177
Pathology 159
Paths 15
Paws 36, 37, 45
PCB's 162, 184
Peering 64
Pelage 34–6
Pelt 29, 30, 163, 168, 179–80, 185–6
  (see also coat)
Penis 140–1
Periodicity in activity 113
Persecution 27, 28, 130, 148
Pesticides 162, 200
Pest status 189–205
Pheasant 75, 193, 195
Phenotypes 169–75
Physiology
  metabolism 44, 57
  reproductive 140
  respiratory 39–43
  sensory 45–56
Phylogeny 18–21
Placenta 146
Play 64
Pleiotropy 174
Plesictis 1
Poikilothermic prey 63, 89–90
Polecat 62, 140, 201
Pollutants
  DDT 162, 200
  heavy metal 162
  mercury 162
  PCBs 162, 184
Poly-chlorinated biphenyls 162, 184
Polygyny 60
Population
  age structure 151–2
  biology 151–62
  censusing 156–8
  density 27, 30, 157, 158
  dynamics 156–8
  regulation of, 188
  size 157
Poultry 193–5, 202

Predation
  on domestic stock 63, 193–5, 202
  on fish 63, 64, 78–9, 84–6, 89–99,
    190–5
  on mammals 80–1
  on salmonids 63, 75, 84–6, 98, 102,
    125, 190, 195, 197
  on seabirds 75–7, 83–4
  on waterfowl 74–6, 79, 82–3, 92, 98,
    190–1, 194
Predatory behaviour 62–74
  aquatic 63–73
  development of 62
  generalist 1, 64
  killing bite 62–3
  prey handling 63, 94–5
  senses used during
    aquatic 46, 73
    terrestrial 45–6
  sequencing of, 46, 62, 64
  specialist 64
  terrestrial
Predators of mink 162, 199
Press reports 2, 187, 188–9, 200
Prey
  availability 64, 79, 95
  capture of, 62
  detection 62
  effects of predation on populations
    80–1, 97–9
  energy content 97–8
  killing 62–3
  remains 14–6
  size 58, 62, 63, 74, 76, 84, 86, 94–6,
    113, 193, 195
  types
    Amphibia 86
    birds 82–4
    Crustacea 87
    domestic stock 193–5
    freshwater fish 84–5, 190
    invertebrates 87–8
    lagomorphs 62, 94, 106, 114 (see
      also hare, rabbit)
    mammals 80–2
    marine fish 85–6
    passerines 83
    reptiles 86–7
    salmonids 63, 75, 84–6, 190,
      195
    seabirds 97, 191–2
    waterfowl 190–1, 194
Promiscuity 60

R-selection 57
Rabbit 62, 63, 64, 71, 75, 77, 80, 81, 88, 90–1, 93, 94–9, 102, 108, 114, 192, 201
Rabbit, as bait 116
Radio-tracking 107, 108, 120–3, 136
  automatic data-logging 122–3
  technique 107, 120–3
  transmitters 120–1
Ranch management 176–8
Ranging behaviour 123–39
Recessive mutations 169, 171
Reintroductions 28
Releases 22, 23, 28, 30, 108, 165, 205
Relationships within Mustelidae 13, 18–21
Reproduction
  age at first breeding 141
  birth 146
  courtship 141–3
  development (also see growth)
    in utero 145–6
    post-partum 146, 148–9
  effect of visual isolation 182–3
  European mink 149
  fecundity 147, 157
  fertility 144, 174
  gestation 60, 99, 140, 145–6, 149
  hormonal effects 140, 143
  implantation
    delayed 140, 145, 146
    effect of day-length 140–1
  induced ovulation 143
  kit development
  lactation 60, 118, 128, 133, 147, 148
  mating 141–3, 158, 175, 176–8
  oestrus 141, 143, 149
  organs 140–1
  ovulation 143, 144
  parturition 118, 146, 147, 148, 177
  seasonal breeding 141, 188
  status 118
  superfoetation 144–5
  superfecundation 144–5, 146, 147
Reproductive system 140–1
Reproductive roles 60
Respiratory biology 39–43
  bradycardia 40–43
  drowning 41–2
  electrocardiogram 40–3
  forced dives 40
  heart rate 40, 43
  oxygen uptake 42
  voluntary dives 40
Respirometry 39–43
Reptiles, in the diet, 86–7
Roundworms 160–2
Russia 28, 29–30
Rut 119, 128, 135, 136, 139, 146, 157

SAGA 179
Salmon 84–6, 190, 195, 197
Salmonids 63, 75, 84–6, 98, 102, 125, 190, 195, 197
Scandinavia 21, 23, 26–8, 176
Scats 14–16
  as home range markers 16
  distinguishing from otter spraints 16
  faecal analysis 74
  identification 15–6
  use in communication 16, 54–5
Scent 54–5
Scent-marking 54–5, 139
  anal scent glands 54–5, 139
  communication 54–55, 139
  faeces 54–5
  postures during, 55
  sign posts 55, 139
  significance of, 139
  urine 54
  ventral scent glands 55
Scotland 25–6
Seabirds 97, 191–2
Seabirds, in the diet 83–4, 75–7, 197
Sea-mink 32–3, 164
Search strategy 66–71
Selective breeding 35, 167–75
Selective pressures 19, 59
Senses
  used during predation 45–6
    on land 45–6
    underwater 46, 73
  hearing 53–4
  olfaction 54–5
  vibrissae 55–6
  vision 46–7, 48–52
    accommodation 50–2
    acuity 46–7
    amphibious vision 46
    on land 45–7
    underwater 46
Sex ratio 151, 155–6
Sex differences
  activity patterns 107–11
  diet 93–6
  habitat utilisation 102–3

Index

Sexual dimorphism
  ecological inplications of, 58–61
  evolution of 58, 96
  of bodyweight 7, 58, 94
  of skull shape 59
  ratio 58
  reproductive roles 60–1
  theories of 58–61
Sight
  colour vision 49–50
  diurnal vision 45, 47, 48–50
  motion detection 47
  nocturnal vision 45–7, 48–50
  structure of the eye 48
    retina 48–9
  visual communication
Signals
  olfactory 54–5
Signs 14–7
Silverblu (colour variety) 4, 167, 170–4
Size 4, 11
  and breeding success 57, 60–1
  at sexual maturity 7–8
Skeleton 8, 9, 12
  appendicular 8–9
  CL⅛ cranial 9–13
Skull 8–11, 12, 19, 20
  comparison of American and
    European mink 12
  of male and female 10, 118
  sexual dimorphism in, 59
  shape 9, 118
  sutres, and use in aging 118, 154
Smell (see olfaction)
  olfactory communication 54–5
    (see also scent marking)
Social behaviour
  contact patterns 141–2
  intersexual interactions 95, 134, 135,
    139, 141–5,
  intra-sexual interactions 134, 135–9
  olfactory communication 54–5
  scent marking 54–5
  social system 188
  vocal communication 52–4
Social instability 130
South America 13, 23
Spacing patterns (see home range,
  territoriality)
Spain 32
Spatial memory 67–8
Speed
  terrestrial 36, 101, 132
  swimming 38–9
Spread of feral populations 23–8
Status 18–33
Stereotyped behaviour 183
Stoat 45, 62, 94, 124, 140, 201
Sub-species of American mink 21–22
Superfecundation 143, 144–5, 146, 147
Surplus killing 73–5, 193, 202
Surveying 15, 25, 27, 156–7
Sweden 26–7
Swimming ability 36–9, 63–7

Tail, length of
Taming 11
Tapeworms 160–2
Taxonomy
Teeth
  canines 10–11, 59, 95, 142, 152
  carnassial 10–11, 95, 152–4
  development of 11
  premolars 11
  techniques for aging 118, 152–4
Testes 140
Telemetry, see also radio-tracking
  of heart rate 40–3, 44–5
Territoriality 123–39
  defence 124, 133–9
  definition 124
  effect of linearity 128–9
  effect on prey populations 99, 201–2
  seasonal variation in 128
  sex-differences 124, 134–9
Thermoregulation 57
Ticks 159–60
Toes 4, 14
Tracking, in snow 156
Tracks 14–5
Transient individuals 130, 137, 143
Trappability 155
Trapping 23, 155, 157–8, 163–6,
  184–5, 202–3
  bait 116
  cage 116
  leg-hold 185
  limitation of, 120
  seasonal effects 118–9
  success 23, 27, 118–20
  techniques 115–20
Tri-hybrid crosses 173
Trout 75, 84, 98, 102, 195

USA 125, 165, 176, 190
USSR 28, 29–30

Underwater vision 46–7
Urinary incontinence 184
Urine 54, 142

Vermin 2
Vibrissae
  structure 55–6
  use during hunting 55–6
Viral diseases
Vision
  colour vision
  structure of the eye 48–52
  structure of the retina 48–50
    cones 48–50
    rods 48–50
    summation 48–9
    tapetum 49
Visual acuity 46–7
  comparative studies 47, 50–1
  effect of submersion on 47, 50–1
  effect of illumination on 46–7
  estimation of 47
Visual isolation 182–3

Visual pathways 50
Vocal
  communication 52–4
  repetoire 52–4
Voles 81, 108

Wading birds 77
Wales 23–5
Waterfowl 74–6, 79, 92, 98, 190–1, 194
Water temperature 63, 89–90
Weaning 149, 177–8
Weasel 45, 53, 62, 64, 94, 124, 140
Webbing 4, 13, 36
Weight 7–8
Welfare 180–4, 185
  guidelines 181–2
Western Europe 26
Whiskers, see vibrissae

Yukon mink 22

Zygomatic arch 10